Sustainable Development Goals Series

The **Sustainable Development Goals Series** is Springer Nature's inaugural cross-imprint book series that addresses and supports the United Nations' seventeen Sustainable Development Goals. The series fosters comprehensive research focused on these global targets and endeavours to address some of society's greatest grand challenges. The SDGs are inherently multidisciplinary, and they bring people working across different fields together and working towards a common goal. In this spirit, the Sustainable Development Goals series is the first at Springer Nature to publish books under both the Springer and Palgrave Macmillan imprints, bringing the strengths of our imprints together.

The Sustainable Development Goals Series is organized into eighteen subseries: one subseries based around each of the seventeen respective Sustainable Development Goals, and an eighteenth subseries, "Connecting the Goals", which serves as a home for volumes addressing multiple goals or studying the SDGs as a whole. Each subseries is guided by an expert Subseries Advisor with years or decades of experience studying and addressing core components of their respective Goal.

The SDG Series has a remit as broad as the SDGs themselves, and contributions are welcome from scientists, academics, policymakers, and researchers working in fields related to any of the seventeen goals. If you are interested in contributing a monograph or curated volume to the series, please contact the Publishers: Zachary Romano [Springer; zachary.romano@springer.com] and Rachael Ballard [Palgrave Macmillan; rachael.ballard@palgrave.com].

Kaitano Dube
Ikechukwu O. Ezeuduji
Magdalena Petronella Swart
Editors

Tourism in Rural Protected Areas

Trends and Opportunities in Southern Africa

Springer

Editors
Kaitano Dube
Department of Tourism and Integrated Communication
Vaal University of Technology
Vanderbijlpark, South Africa

Ikechukwu O. Ezeuduji
Department of Recreation and Tourism
University of Zululand
KwaDlangezwa Campus, Zululand
South Africa

Magdalena Petronella Swart
Department of Applied Management
University of South Africa
Pretoria, South Africa

ISSN 2523-3084 ISSN 2523-3092 (electronic)
Sustainable Development Goals Series
ISBN 978-3-031-87332-4 ISBN 978-3-031-87333-1 (eBook)
https://doi.org/10.1007/978-3-031-87333-1

This Springer imprint is published by the registered company Springer Nature Switzerland AG
The registered company address is: Gewerbestrasse 11, 6330 Cham, Switzerland

Peer Review Process

All the chapters underwent a rigorous double-blind peer review process by expert tourism scholars to ensure the high standards of published work. Double-blind peer review is mandatory for South African-based publishers to fulfil the requirements of Department of Higher Education and Training (DHET) for recognised research outputs for subsidy purposes. Through the review process, some submitted chapters were rejected. Only those chapters accepted by the reviewers and met the standards set by the editorial team are published herein.

Contents

About the Authors

Simiso Lindokuhle Mabaso holds a DPhil in Management Sciences (Hospitality and Tourism) from the Durban University of Technology (DUT), focusing on sustainable tourism economies. He also earned a Master's in Tourism Management (2011) and a B.Tech in Tourism Management. Mabaso has been a Lecturer at DUT since 2014, specialising in entrepreneurial skills, business management, and tourism development. He previously lectured at the Tshwane University of Technology (2011–2013) and DUT (2006–2011). His research has been presented internationally, and he co-authored a paper on tourism recovery post-COVID-19, contributing to the book *Crisis Management, Destination Recovery and Sustainability* (2022).

Chiedza N. Mutanga is a Senior Research Fellow in Sustainable Tourism at the Okavango Research Institute, University of Botswana. She holds a DPhil in Tourism and Wildlife Management from the Chinhoyi University of Technology, Zimbabwe. Her research interests encompass sustainable tourism development, focusing on wildlife tourism, nature-based tourism, community-based tourism, community livelihoods, ecotourism, protected area tourism, community-based natural resource management, and wildlife resource conservation. Chiedza has authored over 40 peer-reviewed scientific publications in these fields.

Joseph E. Mbaiwa is a Professor of Tourism studies at the Okavango Research Institute (ORI), University of Botswana. He holds a PhD in Recreation, Park and Tourism Sciences from Texas A&M University, USA. Joe Mbaiwa has over 20 years of experience in tourism and natural resource management research and consultancy experience in Botswana. He has published extensively on tourism development, rural livelihoods, and conservation issues.

Reniko Gondo holds a PhD degree from the University of Botswana. He earned his MSc. and BSc. degrees from Bindura University of Science Education and the Zimbabwe Open University (Zimbabwe), respectively. He is a Senior Research Fellow at the Okavango Research Institute (ORI), University of Botswana (UB). Gondo has authored over 20 publications focusing on the interface between Indigenous and Western knowledge in the governance of natural resources. He serves as a reviewer for *Water Resources Management*, *Scientific African*, *Sustainable Water Resources Management*, and *Geography and Sustainability*.

Olga L. Kupika is an Associate Professor of Natural Resources Management, a UNFCCC-registered climate change expert, and a UNESCO Earth Network volunteer. She holds a PhD in Natural Resources Management Conservation from the Chinhoyi University of Technology, Zimbabwe. She is currently engaged as a Senior Research Fellow in Climate Change Adaptation at the Okavango Research Institute, University of Botswana. Olga has (co) authored over 45 peer-reviewed scientific publications in line with her research interests, which include climate change, biodiversity conservation, community-based natural resource management, ethnobiology, and natural resources policy. Olga has experience leading and collaborating with multi-disciplinary research teams on climate change, conservation of natural resources, and sustainable livelihoods. She has been awarded the EU-DREAM PhD Scholarship, the Climate Impact Research Capacity and Leadership Enhancement (CIRCLE) Post Masters Fellowship, and the Climate Research for Development (CR4D) Postdoctoral Fellowship, which were implemented by the African Academy of Sciences (AAS).

Ndabezinhle Dube holds a Master of Science in Hospitality and Tourism from Chinhoyi University of Technology in Zimbabwe and a Bachelor of Science Honours in Tourism and Hospitality Management from Lupane State University in Zimbabwe. His research interests include wildlife tourism, propoor tourism, and community-based tourism (CBT).

Ngoni C. Shereni is a Senior Postdoctoral Research Fellow in the School of Tourism and Hospitality at the University of Johannesburg. He also holds a PhD in Tourism and Hospitality from the same university. His research interests include climate change, sustainable tourism, sustainable development goals (SDGs), and community-based natural resource management (CBNRM) practices.

Hugo van den Berg holds a PhD in Environmental Management, which he obtained from the University of South Africa in 2023. Before becoming an academic, he pursued a career in the supply chain environment for several years as a distribution manager and warehouse manager. He started his academic career as a lecturer at the Technikon Witwatersrand. In 2006, Rand Afrikaans University (RAU) merged with Technikon Witwatersrand and became the University of Johannesburg (UJ), where he is in the Department of Transportation and Supply Chain Management. His areas of expertise include logistics, procurement, supply chain management, and warehouse management, and he is actively involved in research methodology and work-integrated learning (WIL) activities. Dr. van den Berg has also served as Acting Deputy Head of the Department. His current research interests lie in sustainable supply chain management and environmental aspects of supply chain management.

Kevin Mearns is a full professor in the Department of Environmental Sciences at the University of South Africa (UNISA). Professor Mearns specialises in the research areas of sustainability, sustainable tourism, community-based tourism, ecotourism, and the social aspects of protected areas in southern and eastern Africa. Professor Mearns has supervised more than 44 postgraduate masters and doctoral candidates to completion and has published more than 74 peer-reviewed scholarly journal articles, chapters, and conference proceedings

Michael Kuseni is a tourism scholar with a Doctorate in Management Sciences, specialising in tourism, and a Master of Technology degree in Tourism and Hospitality Management from Tshwane University of Technology, South Africa. He also has a Bachelor of Commerce Honours degree from Midlands State University in Zimbabwe. He is a Postdoctoral Research Fellow at Tshwane University of Technology. His research focus includes tourism marketing, sustainable tourism, visitor behaviour and management, and tourism in protected areas.

Uwe P. Hermann is an Associate Professor in the Department of Tourism Management at the Tshwane University of Technology, Pretoria. He holds a PhD (Tourism Management) from North-West University and a PhD (Curriculum Development and Instructional Design) from the University of Pretoria. He has over 40 articles in the field of tourism, with a particular focus on tourism in protected areas, niche events, and tourism curriculum development. Professor Uwe heads various committees at the Tshwane University of Technology.

Pearl Siyanda Sifolo is a seasoned tourism expert with strong academic and practical experience. She has a Doctorate in Business Administration, focusing on supply chain management and stakeholder engagement in tourism. Her research interests include digital collaboration in African tourism and youth entrepreneurship. She has published extensively and holds various leadership positions in the travel and tourism industry.

Nomfundo Sinethemba Mzimela is a PhD candidate at the University of Zululand (UNIZULU) South Africa. She holds a Bachelor of Tourism Studies from UNIZULU (2018), Postgraduate Certificate in Education from UNIZULU (2019), a Bachelor of Arts Honours in Recreation and Tourism from UNIZULU (2020), and a Master of Arts in Recreation and Tourism from UNIZULU (2023). She taught Language and Tourism studies as a secondary educator at Empangeni High School (2019–2021). She was also a tutor, coordinator, and mentor at the University of Zululand (2020–2023). She has published two international conference papers around her research interests and contributed to two book chapter publications and journal articles. She also co-supervises postgraduate research students (honours). She is a scholar passionate about community engagement and nature-based tourism sustainability.

Ikechukwu O. Ezeuduji obtained a PhD from the BOKU—University of Natural Resources and Applied Life Sciences, Vienna, Austria. After that, he received a Master of Business Administration (MBA) from The Open University, Milton Keynes, UK, and a Postgraduate Diploma in Higher

Education from the University of KwaZulu-Natal, South Africa. He taught at the University of Johannesburg and the Cape Peninsula University of Technology before joining the University of Zululand. He is a Full Professor of Tourism Management in the Department of Recreation and Tourism, University of Zululand, South Africa. He teaches and supervises postgraduate students (honours, master's, and doctoral).

Sibusiso David Ntshangase has a PhD (Tourism and Recreation) from the University of Zululand, South Africa, where he currently serves as a Lecturer. He is an emerging scholar in the areas of tourism entrepreneurship and tourism in protected areas. He has published around ten journal articles on his research interests and contributed to four international conference papers. He is a scholar passionate about community engagement and sustainable rural development.

Nelisile Z. Mgabhi is a PhD candidate at the University of Zululand (UNIZULU), South Africa. She holds a Bachelor of Tourism Studies from UNIZULU (2015), a Bachelor of Arts Honours in Recreation and Tourism from UNIZULU (2016), a Graduate Certificate in Education from the University of KwaZulu-Natal, South Africa (2017), and a Master of Arts in Recreation and Tourism from UNIZULU (2021). She was a tutor and mentor at UNIZULU (2019–2021) and an Assistant Lecturer at UNIZULU (2022–2023), teaching postgraduate students. She also coordinated a Postgraduate Diploma in Recreation and Tourism (PDRT) programme and co-supervised postgraduate students (honours). She is an emerging researcher whose work has been published in high-impact journals and international conferences. Her research interest lies in rural tourism development.

Part I

Background and Introduction

Evolving Narratives for Tourism in Southern Africa's Rural and Protected Areas

1

Ikechukwu O. Ezeuduji, Magdalena Petronella Swart, and Kaitano Dube

Abstract

Tourists visiting Africa, especially Southern Africa, visit rural and protected areas for various reasons. Since their needs are heterogeneous, it becomes critical that tourism in rural and protected areas should embrace the contingency management concept of "there is no one-size-fits-all approach." This introductory chapter explored the evolving narratives for tourism activities in Southern Africa's rural and protected areas. Key research themes found in this book, including research developments around the study areas of tourism in rural and protected areas, are introduced. It additionally presented an overview of the importance of tourism in rural and protected areas, community participation in rural and protected area-based tourism planning, and strategies for rural and protected area-based tourism planning, in the Southern African context. It established that tourists who visit Southern Africa's rural and protected areas seek nature experiences, nostalgic experiences, novelty, escape, rural landscape, "rural idyll" experiences, and friendly populations. In conclusion, the African *Ubuntu* ideology, based on the moral precept of supporting life via reciprocal care and sharing amongst humans and nonhumans, is one viable sustainability option for the effective development of tourism in rural and protected areas in Southern Africa.

Keywords

Rural tourism · Tourism in protected areas · Ecotourism · *Ubuntu* · Southern Africa

I. O. Ezeuduji (✉)
Department of Recreation and Tourism, University of Zululand, KwaDlangezwa Campus, Zululand, South Africa
e-mail: ezeudujii@unizulu.ac.za

M. P. Swart
Department of Applied Management, University of South Africa, Pretoria, South Africa
e-mail: Swartmp@unisa.ac.za

K. Dube
Department of Tourism and Integrated Communication, Vaal University of Technology, Vanderbijlpark, South Africa

Emirates Aviation University, Dubai Academic City, Dubai, United Arab Emirates
e-mail: kaitanod@vut.ac.za

1.1 Introduction

Many tourism activities are focused on nature. Nature-based tourism relies so much on protected areas, widely found in rural areas. In 2023, The Global Economy (2024) reported that the global rural population was estimated at 38.64%, with Papua New Guinea having the highest value of 86.28%, and Bermuda, having the lowest value of

K. Dube et al. (eds.), *Tourism in Rural Protected Areas*, Sustainable Development Goals Series,
https://doi.org/10.1007/978-3-031-87333-1_1

0%. From the same source (The Global Economy, 2024), the rural population in Africa, in 2023, was estimated at 51.28%, with Burundi having the highest share of 85.22% and Gabon, having the lowest share of 8.97%. Hence, rural populations remain significant, not just in the African perspective, but also in the global context, and tourism in rural and protected areas is seen as a catalyst for local economic development, and in ultimately achieving the United Nations' sustainable development goals. Hence, the global interest in rural tourism studies (see Figs. 1.1 and 1.2) is growing.

Recent studies around protected areas and nature-based tourism have used themes such as "ecotourism," "nature conservation, biodiversity, and sustainability," "national parks, recreation, and climate change," "sustainable management and development," "sustainable tourism," and "geotourism and rural tourism" (Silva et al., 2023). Rural tourism (tourism in rural spaces) and tourism in protected areas (geographical areas dedicated to biodiversity conservation) have strong focus on nature, people, and education. Hence, the concept of ecotourism may be commonly used to integrate these related tourism types. The International Ecotourism Society (2015) defined ecotourism as "responsible travel to natural areas that conserves the environment, sustains the well-being of the local people, and involves interpretation and education." Because the local communities, not just visitors, are subjects of tourism in rural and protected areas, local participation is therefore greatly encouraged in rural and protected areas tourism planning (Bello et al., 2017).

Many of the early conservation initiatives in Southern Africa at the later stage of 1800s and at the early periods of 1900s either uprooted indigenous populations or limited indigenous people's access to their natural resources (Snyman, 2014a). This incidentally changed how the indigenous communities felt about the protected areas, but later, attempts were made to ease the mounting conflict situations. In an attempt to win over communities, conservation and ecotourism approaches emerged in the final decades of the twentieth century, incorporating them more and

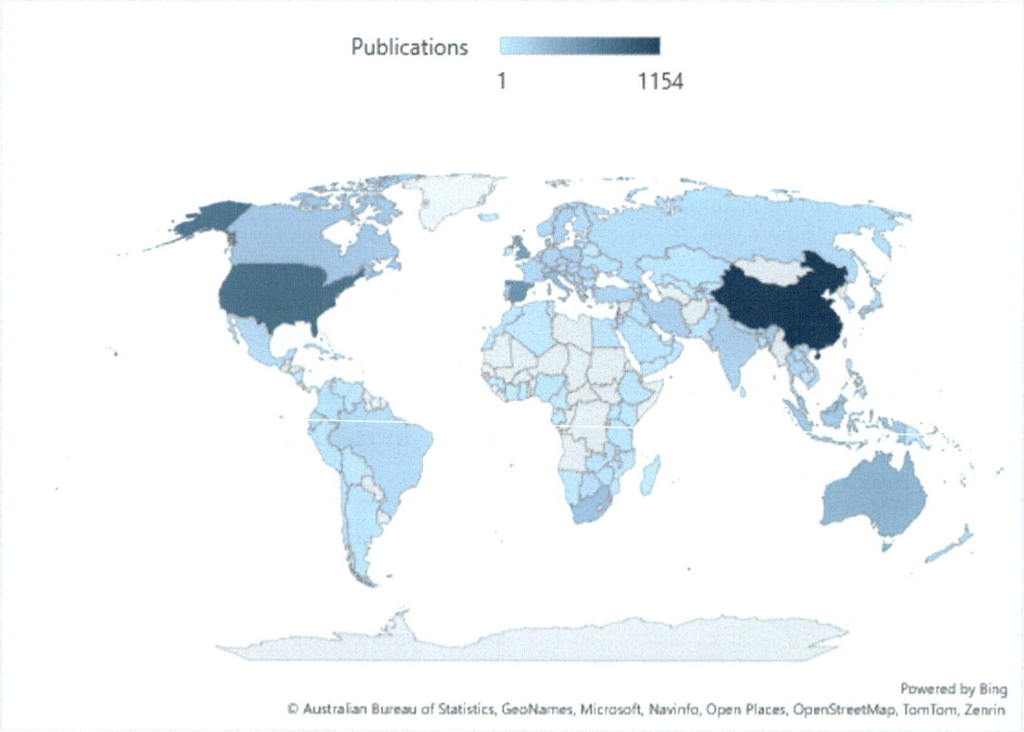

Fig. 1.1 Global distribution of interest in rural tourism studies between 1980 and 2024: Data from Scopus

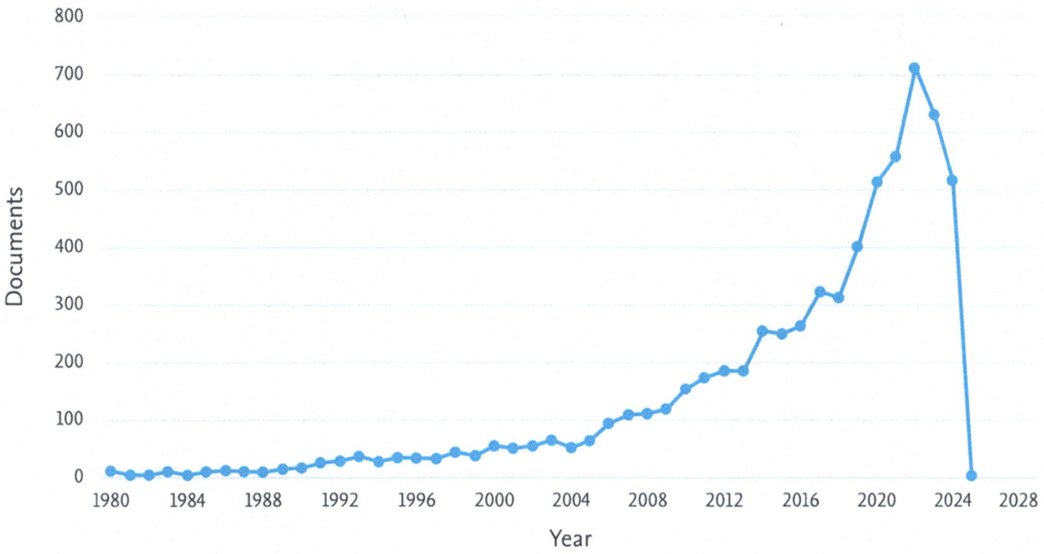

Fig. 1.2 Trends in publications on rural tourism between 1980 and 2024: Source: Scopus Data

more in the decision-making and benefit-sharing processes. Notwithstanding the uneven outcomes of these initiatives, it is obvious that the attitudes and actions of the populations residing in or close to protected areas will have a significant impact on the effectiveness of conservation efforts in the future and, by extension, ecotourism in many regions (Snyman, 2014a).

This introductory chapter, focusing on Southern Africa (where many countries depend on tourism as a major contributor to the Gross Domestic Product), presents the evolving narratives for tourism in rural and protected areas. Figures 1.3, 1.4, and 1.5 show self-explanatory, regarding publishers of rural tourism studies, themes in rural tourism studies, and the evolution of in-country focus on rural tourism studies, respectively.

1.2 Why Is Tourism in Rural and Protected Areas Important?

A study by Snyman (2012) evaluated the various effects of employment related to ecotourism in Botswana, Malawi, and Namibia. Due to ecotourism operations, rural communities are becoming more integrated into the market economy, according to analysis of household income, social welfare implications, and the number of persons indirectly impacted by ecotourism employment. Monthly revenue from jobs in ecotourism enables households to invest in assets, education, and "luxury" commodities, improving social welfare and financial security in isolated, rural locations. Ecotourism employment, hence, has a favorable impact on attitudes toward conservation and tourism (Snyman, 2012, 2014b). According to Munthali (2007), many Southern African nations are implementing policies that support the combination of rural development with biodiversity conservation to lessen rural poverty.

In emerging Southern African nations, which have sizable rural populations, but with limited economic activity, rural development is very crucial. Therefore, the popularity of rural tourism is rising as a substitute for urban tourism (Ezeuduji & Mgabhi, 2023). The growth of the rural tourism industry helps the socioeconomic development of emerging nations. In Southern Africa, rural communities face major obstacles to growth, and little possibilities. Thus, for rural tourism to be a useful tool in the fight against poverty, its goods must be both sustainable and commercially viable (Haywood et al., 2020).

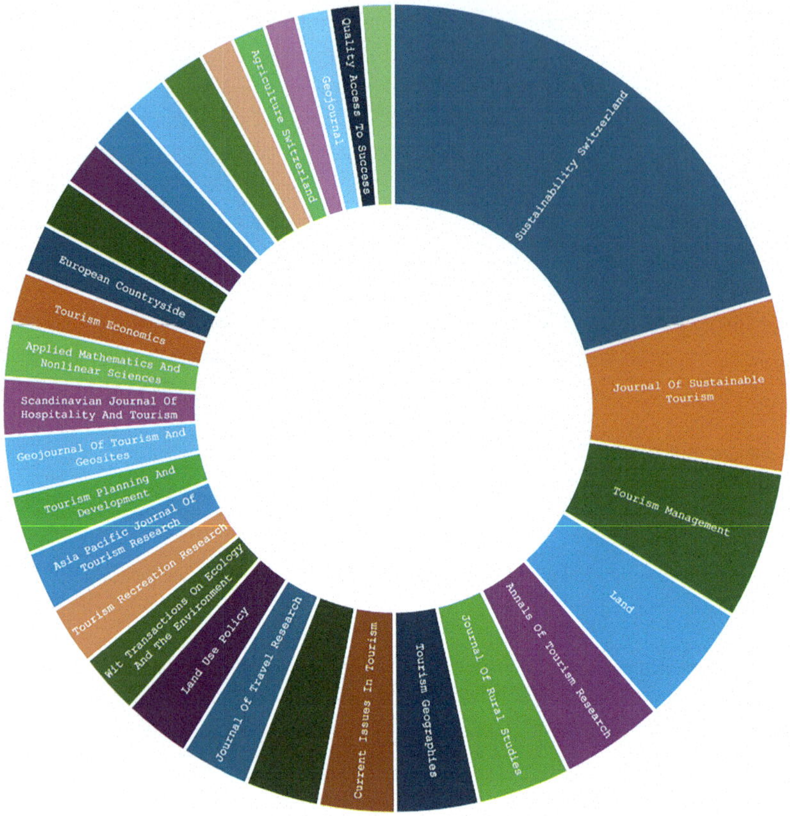

Fig. 1.3 Top 30 publishers of rural tourism studies: Data Based on Scopus Data Base

One way of integrating tourism in rural and protected areas is to increase the critical mass of visitor attractions within and around protected areas, where the local communities surrounding this protected area can create alternative rural tourism goods and services for incoming visitors (Ezeuduji & Mgabhi, 2023). Lulu et al. (2024) conducted a bibliometric analysis of residents' views regarding rural tourism development, and one of the findings is that the focus of research has gradually moved from first examining how locals perceive changes in the conservation area to a community-based viewpoint and ultimately an investigation of locals' happiness and quality of life.

Sustainable tourism development in rural and protected areas must strive toward developing a strong brand that will make such development highly competitive in the global tourism industry. The brand essence must be attractive in the long term and have a clear targeted audience, sizeable enough to sustain the development. The brand essence of rural and protected areas tourism in Southern Africa consists of its functional and emotional benefits. Ezeuduji (2015a) posits that the friendly people, varied natural and cultural features, favorable climate, and distinctive rural landscape of sub-Saharan Africa all contribute to the region's rural tourism industry's functional benefits. The derived emotional benefits lie in the "feel-good" aspect that arises whenever visitors purchase local goods and souvenirs or are treated to "rural idyll" experiences, that are unmatched on other continents or subcontinents.

According to Moswete et al. (2020), protected areas are important to the country and have grown

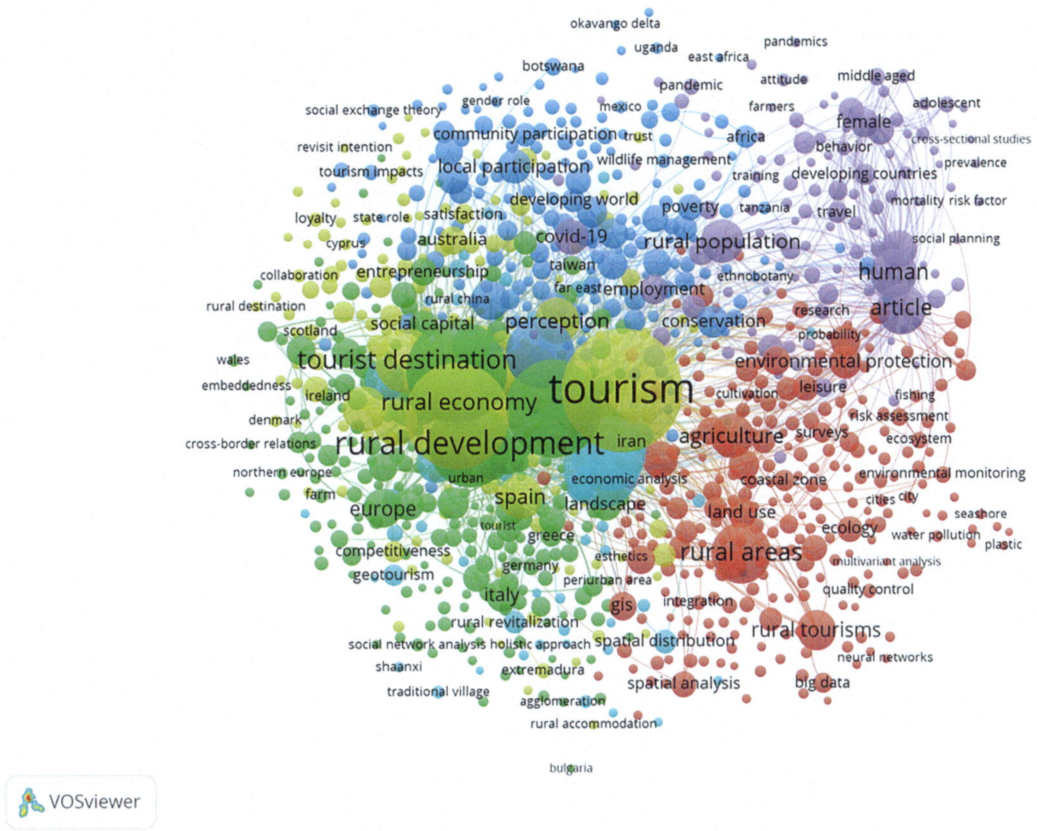

Fig. 1.4 Key themes in rural tourism studies

to be sources of advantages, however, in other cases, they have caused disputes between various parties, such as park officials and people living in nearby localities. Ezeuduji and Rid (2011), therefore called for contingency (tailored to the context) and effective stakeholder management, clarity of functions, and benefits in rural and protected areas tourism development. Also, Snyman (2014a) posits that more effective, fair, and long-lasting community-based conservation and eco-tourism approaches will result from managing and comprehending expectations and attitudes within the community under different socioeconomic conditions. According to Snyman (2014a, 2014b), Southern African communities that are close to or within protected areas, in general, recognize the value of conservation and rural tourism.

1.3 Community Participation in Rural and Protected Area-Based Tourism Planning

It is commonly recognized that poor rural residents in less developed nations may experience difficulties due to tourism. One of the main reasons rural communities struggle to escape the cycle of poverty is a lack of access to pertinent information. Survivalist communities need to be trained and educated to mobilize themselves economically, possibly through the growth of local tourism (Fairer-Wessels, 2017). Western scientific theories of conserving nature have focused on isolating it from human lifestyles, particularly through the creation of "protected areas." Hence, Mabele et al. (2022, p. 92) reported that an alter-

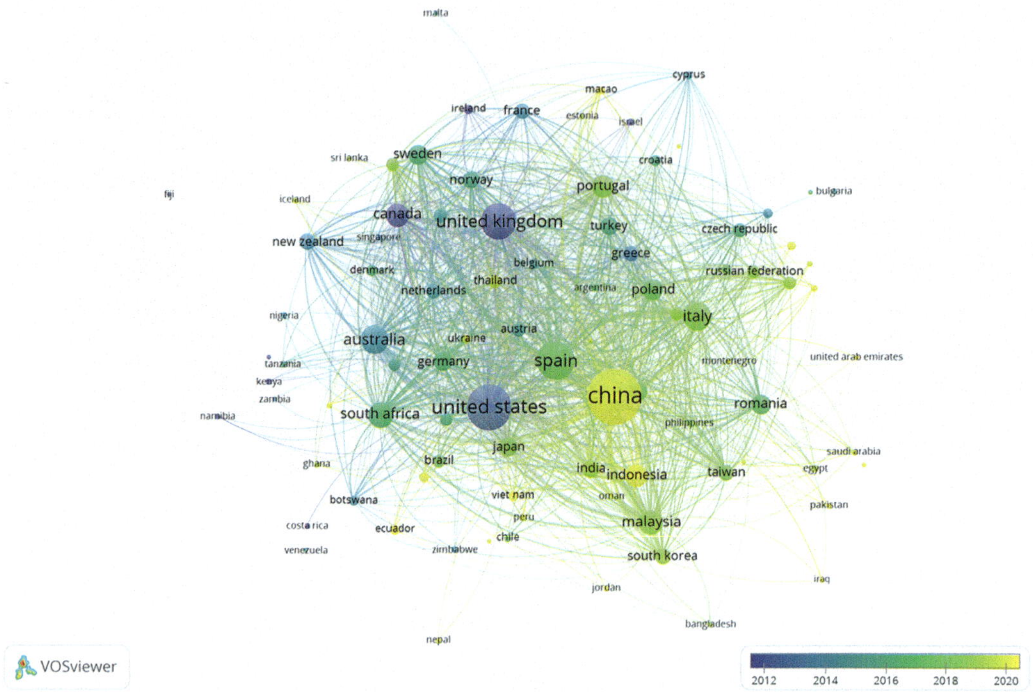

Fig. 1.5 Evolution of in-country focus on rural tourism studies over time

native concept known as "promoted areas" has been put forth under the umbrella of "convivial conservation," which supports socio-ecological justice and structural changes in the international economic system. This concept calls for conservation that promotes nature for, to, and by humanity.

Despite decades of re-examination, nature conservation, particularly in Africa, is still struggling to break free from the customs of its colonial past. Currently, residents are unable to fully engage in the planning and design of the conservation efforts where they reside, which limits the potential benefits for residents and frequently restricts the ability of Indigenous people to conserve. This is due to the persistent coloniality of conservation, which is more technocrat-oriented. This disturbs the equitable balance that may exist between nature conservation and the socioeconomic advancement of nearby or local communities (Bourgeois et al., 2023). For example, in Southern Africa, wildlife safaris are synonymous with tourism. Establishing so-called peace parks that cross state borders has been important in the post-colonial era, not just for boosting tourism in Southern Africa but also for modernizing conservation regulations and boosting rural economies (Ferreira, 2004). However, the conservation of wildlife and tourism are adversely impacted by the uncertain political conditions in certain nations.

Mabele et al. (2022) contend that decolonial ideas in conservation science and practice are most appropriate for "promoted areas." They posit that the African *Ubuntu* ideology, which is based on the moral precept of supporting life via reciprocal care and sharing amongst and among humans and nonhumans, is one viable "decolonial option" in Southern Africa. *Ubuntu*, which emphasizes the numerous connections between humans and nonhumans, detaches itself from Western conceptions of human–environment relations. Therefore, Mabele et al. believe that *Ubuntu's* "promoted areas" restore harmony between humans and the natural world by discouraging individualistic and excessive extraction of nonhuman nature and promoting relationships between humans and nonhumans, based on respect, solidarity, and cooperation.

In sub-Saharan Africa, nature-based tourism and community-based natural resource management frequently work together to promote economic growth and conservation (Gronau et al., 2017). Recent studies, conducted in Southern Africa (Bello et al., 2017; Ezeuduji & Mgabhi, 2023), have identified several barriers to community involvement in rural and protected areas tourism planning. These include apathy; insufficient funding; incomplete information; low levels of education; unequal benefit distribution; a lack of skilled labor; a centralized public administration; a lack of coordination; and conflicts between people and wildlife. The results point to the necessity of increasing support for community involvement as well as the inclusion of tactics that can encourage community involvement, suited to the circumstances of emerging nations. Mgabhi and Ezeuduji (2023), whose study was conducted in South Africa, therefore noted that the development of human resources would be required for the successful development of rural tourism. Other critical needs include infrastructural development and leadership commitment in driving the process of rural tourism development through capacity building.

Another form of community participation in rural and protected area-based tourism planning practiced in several Southern African countries, including Zimbabwe, Namibia, Botswana, Zambia, and Mozambique is Community-Based Natural Resource Management (CBNRM), an institution aimed at promoting and empowering communal people with economic and social developments through sustainable use of wildlife (Simuchimba et al., 2024).

1.4 Strategies for Rural and Protected Area-Based Tourism Planning: Demand and Supply Perspectives

The study by Rid et al. (2014) stated that to encourage the growth of rural tourism, more research is needed to determine the factors that drive travelers to participate in this tourism market sector. Their (Rid et al., 2014) rural tourism market segmentation study in The Gambia highlighted "heritage and nature seekers," "multi-experiences seekers," "multi-experiences and beach Seekers," and "sun and beach seekers" as the four different rural tourist segments. In the same vein, Lwoga and Maturo (2020), whose study was conducted around Kilimanjaro Mountain, situated in Tanzania, examined the reasons for rural tourism and divided the market according to push and pull motivational factors. Their study (Lwoga and Maturo) offered concrete proof that the desire for a more traditional rural lifestyle drives the motivation of rural visitors in African villages. Other reasons include unwinding with loved ones, discovering regional farming practices, interacting with the locals, taking in the scenery, and giving back to the community. Four segments of "multi-experiences," "authenticity-learners," "unwinding with friends and family," "visiting farms and nature enthusiasts," and other informal segments were produced by clustering the motivations. Hence, Lwoga and Maturo's study makes the case that there are many different types of rural tourists.

As suggested by Ezeuduji (2015a, p. 34), a brand position statement for rural tourism destinations in sub-Saharan Africa could read: "To tourists seeking nostalgic experiences, novelty, and escape, rural sub-Saharan Africa offers authentic rural landscape, 'rural idyll' experience, and amazing friendly populations." In addition, "nature experiences" are predominantly sought-after by tourists who visit protected areas in sub-Saharan Africa, including Southern Africa. For rural tourism demand to meet its supply, the products and services offered for rural tourism must match its demand. The dynamic capability development from the rural population, to offer the required products and services becomes critical for success. However, low skill levels, a strong reliance on natural resources, high rates of unemployment, and poverty, are the main characteristics of rural African communities. Natural resources are under strain due to growing populations and the effects of climate change, making the topic of sustainable land use increasingly crucial (Snyman, 2014b). Economic diversification strategies are nonetheless needed, so far, the challenges of rural tourism development (from the infrastructure and community preparedness and capability perspectives), as outlined in the previous section, are continuously addressed. A previous study by Rid et al. (2014) posits that rural tourism is often mentioned as a diversification tactic that could spur local economic development in outlying regions.

It takes tact to manage a local community's shift from traditional sources of income to rural tourism (Ezeuduji, 2017). In many African nations, there is a danger that rural tourism development could make locals objects rather than subjects, and as a result, they may not directly benefit from tourism (see Ezeuduji & Rid, 2011). The authors (p. 187) therefore suggested that the growth of rural tourism will need "a clear value position, operational excellence and strong customer relations to gain a market leadership" to provide destinations with the sought-after competitive edge. This finding should be utilized to provide support to the rural tourism stakeholders that put communities' needs first, to be able to manage this development proactively and to keep tourism-related income in rural areas. Planning and showcasing activities that would provide distinctive rural tourism experiences to the tourists, may help rural areas to establish a brand identity in the rural tourism industry. According to Ezeuduji (2023), art and handicrafts, religious rituals, hunting and farming traditions, cuisine, music, dance, family dynamics, village celebrations, cultural performances, and many other aspects of the living legacy of people's inheritances from their ancestors are just a few examples of the diverse ways that Africa's heritage manifests itself.

Rid et al. (2014) proposed a development route that might support the diversification of the tourism industry. They put forward that activities for event-based rural tourism that match the goals of the identified rural tourism market segments would be part of the growth route; as these activities might help rural communities to be well engaged in rural tourism planning and development, hence lowering rates of economic leakage.

The study by Ezeuduji (2015b) presented a generic competitive strategy that rural sub-Saharan Africa can use to get an advantage in the global tourism market. The strategy decision to focus on event-based rural tourism was made after the analysis of the subcontinent's rural tourism resources and capabilities, critical success determinants for the sector, the external environment surrounding the growth of rural tourism, and the local institutional stakeholder groups. This event-based tourism was selected as one of the emerging strategies that could help strike a balance between the need to use top-down resources (national governments and not-for-profit organizations as partners in terms of business training and initial financing) and "bottom-up" values (comprising local values and ownership). In implementing this strategy, the residents can host events for visitors, which will result in the establishment of unique rural tourism experiences, while protecting the distinctive natural and cultural legacy that this subcontinent is blessed with.

1.5 Conclusions

This introductory chapter established that tourists who visit Southern Africa's rural and protected areas are heterogeneous and seek nature experiences, nostalgic experiences, novelty, escape, rural landscape, "rural idyll" experiences, and friendly populations. African Ubuntu ideology, based on the moral precept of supporting life via reciprocal care and sharing amongst humans and nonhumans, is one viable sustainability option for the effective development of tourism in rural and protected areas in Southern Africa.

Lane and Kastenholz (2015, p. 1133) advocated for the research-driven establishment of a "New-Generation Rural Tourism," founded on professional approaches to holistic and sustainability-enhancing management, improved understanding of markets and contemporary marketing, and informed destination planning and management. It is recommended that an international research group on rural tourism be formed to investigate, support, and evaluate new-generation rural tourism. The subsequent chapters in this book play their parts in addressing these research needs regarding tourism in rural and protected areas, focusing on Southern African case studies.

References

Bello, F. G., Lovelock, B., & Carr, N. (2017). Constraints of community participation in protected area-based tourism planning: the case of Malawi. *Journal of Ecotourism, 16*(2), 131–151. https://doi.org/10.1080/14724049.2016.1251444

Bourgeois, R., Guerbois, C., Giva, N., Mugabe, P., Mukamuri, B., Fynn, R., Daré, W., Motsholapheko, M., Nare, L., Delay, E., Ducrot, R., Bucuane, J., Mercandalli, S., Le Page, C., & Caron, A. (2023). Using anticipation to unveil drivers of local livelihoods in transfrontier conservation areas: A call for more environmental justice. *People and Nature, 5*(2), 726–741. https://doi.org/10.1002/pan3.10446

Ezeuduji, I. O. (2015a). Brand positioning for sub-Saharan Africa's rural tourism development. *Journal of Economics and Behavioral Studies, 7*(6), 34–41. https://doi.org/10.22610/jebs.v7i6(J).616

Ezeuduji, I. O. (2015b). Strategic event-based rural tourism development for sub-Saharan Africa. *Current Issues in Tourism, 18*(3), 212–228. https://doi.org/10.1080/13683500.2013.787049

Ezeuduji, I. O. (2017). Change management for sub-Saharan Africa's rural tourism development. *Current Issues in Tourism, 20*(9), 946–959. https://doi.org/10.1080/13683500.2014.946892

Ezeuduji, I. O. (2023). Cultural events and tourism in Africa. In D. J. Timothy (Ed.), *Cultural heritage and tourism in Africa* (1st ed., pp. 67–84). Routledge. https://doi.org/10.4324/9781003153955-4

Ezeuduji, I. O., & Mgabhi, N. Z. (2023). Potential for rural tourism development around Hluhluwe iMfolozi Park, South Africa: A local community perspective. *African Journal of Development Studies, 13*(1), 89–113. https://doi.org/10.31920/2634-3649/2023/v13n1a5

Ezeuduji, I. O., & Rid, W. (2011). Rural tourism offer and local community participation in The Gambia. *TOURISMOS: An International Multidisciplinary Journal of Tourism, 6*(2), 187–211.

Fairer-Wessels, F. A. (2017). Determining the impact of information on rural livelihoods and sustainable tourism development near protected areas in Kwa-Zulu Natal, South Africa. *Journal of Sustainable Tourism, 25*(1), 10–25. https://doi.org/10.1080/09669582.2016.1165234

Ferreira, S. (2004). Problems associated with tourism development in Southern Africa: The case of transfrontier conservation areas. *GeoJournal, 60*(3), 301–310. https://doi.org/10.1023/B:GEJO.0000034736.23918.05

Gronau, S., Winter, E., & Grote, U. (2017). Modelling nature-based tourism impacts on rural development and conservation in Sikunga Conservancy, Namibia. *Development Southern Africa, 34*(3), 276–294. https://doi.org/10.1080/0376835X.2016.1269638

Haywood, L. K., Nortje, K., Dafuleya, G., Nethengwe, T., & Sumbana, F. (2020). An assessment for enhancing sustainability in rural tourism products in South Africa. *Development Southern Africa, 37*, 1033–1050. https://doi.org/10.1080/0376835X.2020.1806784

Lane, B., & Kastenholz, E. (2015). Rural tourism: The evolution of practice and research approaches - Towards a new generation concept? *Journal of Sustainable Tourism, 23*(8-9), 1133–1156. https://doi.org/10.1080/09669582.2015.1083997

Lulu, L., Ramachandran, S., Bidin, S., Subramaniam, T., & Chaoyi, C. (2024). A bibliometric analysis of residents' perceptions in rural tourism development using CiteSpace. *Tourism Planning & Development, 21*(4), 438–461. https://doi.org/10.1080/21568316.2023.2209057

Lwoga, N. B., & Maturo, E. (2020). Motivation-based segmentation of rural tourism market in African villages. *Development Southern Africa, 37*(5), 773–790. https://doi.org/10.1080/0376835X.2020.1760791

Mabele, M. B., Krauss, J. E., & Kiwango, W. (2022). Going back to the roots: Ubuntu and just conservation in Southern Africa. *Conservation and Society, 20*(2), 92–102. https://doi.org/10.4103/cs.cs_33_21

Mgabhi, N. Z., & Ezeuduji, I. O. (2023). Stakeholder perceptions of the potentials for rural tourism development around Hluhluwe iMfolozi Park in South Africa. *Journal of Tourism and Development, 42*, 45–65. https://doi.org/10.34624/rtd.v42i0.32664

Moswete, N., Thapa, B., & Darley, W. K. (2020). Local communities' attitudes and support towards the Kgalagadi Transfrontier Park in southwest Botswana. *Sustainability, 12*(4), 1524. https://doi.org/10.3390/su12041524

Munthali, S. M. (2007). Transfrontier conservation areas: Integrating biodiversity and poverty alleviation in Southern Africa. *Natural Resources Forum, 31*(1), 51–60. https://doi.org/10.1111/j.1477-8947.2007.00130.x

Rid, W., Ezeuduji, I. O., & Pröbstl-Haider, U. (2014). Segmentation by motivation for rural tourism activities in The Gambia. *Tourism Management, 40*, 102–116. https://doi.org/10.1016/j.tourman.2013.05.006

Silva, S., Silva, L. F., & Vieira, A. (2023). Protected areas and nature-based tourism: A 30-year bibliometric review. *Sustainability, 15*(15), 11698. https://doi.org/10.3390/su151511698

Simuchimba, G., Blignaut, J. N., & Sjaastad, E. (2024). Estimating the economic multipliers for wildlife-based tourism in Zambia. *Development Southern Africa, 41*, 987–1010. https://doi.org/10.1080/0376835X.2024.2357301

Snyman, S. (2014a). Assessment of the main factors impacting community members' attitudes towards tourism and protected areas in six Southern African countries. *Koedoe, 56*(2), 1–1. https://doi.org/10.4102/koedoe.v56i2.1139

Snyman, S. (2014b). The impact of ecotourism employment on rural household incomes and social welfare in six Southern African countries. *Tourism and Hospitality Research, 14*(1/2), 37–52.

Snyman, S. L. (2012). The role of tourism employment in poverty reduction and community perceptions of conservation and tourism in Southern Africa. *Journal of Sustainable Tourism, 20*(3), 395–416. https://doi.org/10.1080/09669582.2012.657202

The Global Economy. (2024). Rural population, percent - Country rankings. Retrieved August 22, 2024, from https://www.theglobaleconomy.com/rankings/rural_population_percent/

The International Ecotourism Society. (2015). What is ecotourism? Retrieved August 22, 2024, from https://ecotourism.org/what-is-ecotourism/

Part II

Transformation and Sustainability in Rural Tourism Space

2 The Transformative Power of Social Entrepreneurship in Tourism: Practical Implications for Sustainable Development and Social Cohesion

Simiso Lindokuhle Mabaso and Dumsile Cynthia Hlengwa

Abstract

Background: Social entrepreneurship (SE) fosters social cohesion, enhances social capital, and ensures the sustainability of small ventures. SE, especially in the field of tourism, promotes sustainable development because it empowers local communities and serves as a valuable tool for policymakers, entrepreneurs, and organizations seeking to understand and leverage the power of actor networks. This chapter sheds light on the challenges faced by tourism social entrepreneurs and highlights the positive outcomes they achieve through innovative approaches.

Objective: To examine the role of tourism social entrepreneurship in sustainable local development and impact. Methods: Data were collected from 35 tourism-registered business respondents using open-ended questions. Content and thematic techniques were used to analyze the data.

Results: The findings highlight the growing recognition of social entrepreneurship as a business model among tourism practitioners because it prioritizes the creation of actor networks as driving forces for social change. This chapter contributes to the existing literature by providing empirical evidence of the significance of social entrepreneurship in the field of tourism in addressing societal and small venture challenges in the quest for sustainability and impact.

Keywords

Social entrepreneurship · Tourism · Business challenges · Sustainable development · Local communities

2.1 Introduction

Tourism is actively promoted by various governments and development agencies (Khan et al., 2020). This underscores the overarching objectives shared by social entrepreneurship and tourism, which encompass addressing societal issues while maximizing benefits and minimizing the adverse effects of tourism on host communities. These objectives revolve around eradicating socioeconomic problems through the creation of social value ecosystems, aligning with social entrepreneurship practices that prioritize the formation of both social and financial assets (Portales & Portales, 2019).

S. L. Mabaso · D. C. Hlengwa (✉)
Department of Ecotourism, Durban University of Technology, Pietermaritzburg Campus, Pietermaritzburg, South Africa
e-mail: simisom@dut.ac.za; dumisileh@dut.ac.za

K. Dube et al. (eds.), *Tourism in Rural Protected Areas*, Sustainable Development Goals Series,
https://doi.org/10.1007/978-3-031-87333-1_2

Aquino (2022) emphasize that social entrepreneurship serves as a market-driven strategy designed to positively transform communities and society to foster sustainable tourism. Recognizing that community development has long been the cornerstone of the tourism industry, various alternative and community-centric approaches to tourism development have been advocated worldwide to promote sustainability (Snihur et al., 2018). These initiatives highlight the potential of Tourism Social Entrepreneurship (TSE) to drive social change and transformation at the community level.

This study advocates the utilization of Tourism Social Entrepreneurship (TSE) as a strategic mechanism for fostering sustainable community development within destination areas. Specifically focusing on Bergville, a rural locality within the KwaZulu-Natal region of South Africa, the research presented in this chapter examine the applicability of social entrepreneurship in addressing the distinctive challenges and opportunities inherent in such environments. Through a comprehensive analysis of the TSE initiatives implemented in Bergville, the objective is to elucidate the theoretical underpinnings that substantiate their efficacy, while also delineating their practical implications for sustainable development.

In seeking to advance scholarly discourse, this research endeavors to enrich the existing literature by providing empirical evidence elucidating the pivotal role of social entrepreneurship in facilitating sustainable community development within destination locales. By delineating the specific theoretical lacunae addressed by the inquiry and explicating how the study contributes to bridging this gap, the aim is to underscore the significance of comprehending and advocating for social entrepreneurship as an instrumental catalyst for fostering sustainable progress.

2.2 Literature Review

Tourism social entrepreneurship represents a burgeoning field within the broader spectrum of social entrepreneurship, where the principles of business innovation are harnessed to address social challenges within the dynamic context of the tourism industry (De Lange & Dodds, 2017). Rooted in the transformative ethos of social entrepreneurship, TSE emerges as a proactive response to societal needs, aiming to redefine the role of businesses in tourism while prioritizing social impact over traditional profit-centric objectives (Banks, 1987). This literature review delves into foundational concepts, theoretical frameworks, and empirical studies, offering a thorough examination of Tourism Social Entrepreneurship (TSE). It aims to clarify the theoretical foundations of the TSE and its real-world applications in the tourism sector. As we navigate through the evolving landscape of TSE, this review aims to illuminate the key themes and contributions that shape our understanding of the dynamic interplay between social entrepreneurship and the intricacies of the tourism industry.

Social enterprises often begin locally but can expand globally (Hockerts et al., 2022). Celebi et al. (2020) argued that SE play a critical role in improving society's well-being. Radic et al. (2020) conducted a study in Croatia, suggesting that SE could serve as a vital vehicle for sustainable destination development, product diversification, and long-term environmental, social, and economic sustainability.

Some research underscores the continued relevance of Banks and Grambs' conceptualization of social movements. For example, Snow et al. (2014) and Della Porta and Diani (2019) emphasize the role of collective identity and collective action frames in mobilizing individuals and sustaining social movements over time. These works highlight the importance of shared grievances and values in uniting diverse groups around common goals, echoing the collective recognition of social issues.

Contemporary scholarship has shed light on the diverse strategies and goals of social movements. For instance, Jasper (2014) discussed the distinction between innovative, conservative, reactionary, reform, and revolutionary movements. Recent examples, such as the climate justice movement and Black Lives Matter, illustrate the varirious approaches taken by social movements to address environmental degradation and systemic racism, respectively.

Recent studies have highlighted the intersectionality of social movements and complexities of social change. As articulated by Crenshaw (1989) and Collins (2019), intersectional frameworks emphasize the interconnectedness of social identities and systems of oppression, underscoring the importance of addressing multiple axes of inequality within social movements. This perspective recognizes the multifaceted nature of social issues and the diverse strategies employed by movements to enact change.

Kirzner (1999) defined entrepreneurship as the process of identifying discrepancies between specific needs and available resources, establishing resourceful enterprises, pursuing business growth, and continually seeking opportunities for innovation and rational outcomes. SE offers a unique perspective on entrepreneurship by introducing new management models that yield different results (Dahlin et al., 2018). According to Clark et al. (2018), social entrepreneurs focus on addressing market failures, not just those related to price fluctuations or individual access to services or goods, but also market collapses related to the scope of products, innovation, and fair distribution. Littlewood and Holt (2018) asserted that SE and innovation are intricately linked, serving as mechanisms to create and implement solutions for sustainable development challenges. Drucker (2001) differentiates SE from other entrepreneurial endeavors by its emphasis on achieving a social purpose, with both conventional enterprises and SE holding collective value. In SE, the overt and primary drivers are social value, encompassing social businesses, social enterprise capital, and social commitment (Robinson 2006).

Mody and Day (2014) underscore entrepreneurship's critical role in tourism development, both regionally and globally, contributing to the GDP, reducing social challenges through increased fiscal growth, and promoting overall national development. Tourism entrepreneurs transform leisure and recreational opportunities into productive and sustainable growth by generating employment opportunities. The widespread dissemination of tourism products and significant returns can be achieved through social enterprises or cooperatives, granting operators access to markets and economies of scale to compete with large corporations.

Globalization has significantly influenced the international landscape of social entrepreneurship. Enhanced connectivity and information sharing enable social entrepreneurs to learn from and collaborate with their peers globally (Mair & Marti, 2019). Dacin et al. (2010) emphasized the importance of cross-cultural understanding and adapting business models to local contexts for an effective impact. SE has gained international prominence as a means of promoting sustainable development and addressing social issues (Dees, 2001; Mair & Marti, 2006; Nicholls & Cho, 2006). SE employs innovative approaches to create social value, including novel business models, partnerships, and community engagement (Santos, 2012). Austin et al. (2006) argue that SE has the potential to transform traditional economic models and tackle complex social challenges, such as poverty, inequality, and environmental degradation.

Governments play a pivotal role in supporting SE and fostering an environment conducive to their growth (Mair & Marti, 2006; Nicholls & Cho, 2006). However, government policies and regulations can also pose challenges for SE, particularly in unpredictable and inconsistent regulatory environments, often found in developing countries (Dorado, 2006). Successful SE initiatives have emerged globally, including Colombia, Brazil, India, and South Africa, highlighting the potential of SE to create positive social and economic impacts (Austin et al., 2006). Government policies and institutional support remain crucial in fostering a favorable environment for SE to thrive (Saiz-Alvarez et al., 2021). Countries with supportive legal frameworks, financial incentives, and resource access tend to have more vibrant SE ecosystems (Defourny & Nyssens,

2010). Peredo and McLean (2006) noted that governments and institutions can offer tailored funding opportunities for social entrepreneurs, enabling them to access capital for startups and expansion. Financial support can be instrumental in creating innovative solutions to social problems. Access to resources, such as training, mentoring, and networking, is essential for growth and sustainability of social enterprises (Kim et al., 2020). Government institutions can establish programs that offer capacity-building workshops, mentoring schemes, and access to business development services. Such initiatives can enhance skills, knowledge, and chances of success of social entrepreneurs.

SE is gaining traction in the tourism industry as a means of promoting sustainable development and community empowerment (Peredo & McLean, 2006). In particular, hotels have the potential to contribute by engaging in activities that benefit local communities, such as sourcing products and services locally, and providing employment and training opportunities (Kanellou & Spathis, 2013; Kibicho & Kimenyi, 2014). However, SE face challenges such as limited access to capital, lack of supportive infrastructure, and limited networks (Santos, 2012). Additionally, measuring social impact and achieving financial sustainability can be challenging because SE often involves addressing complex and long-term social issues (Dees, 2001; Mair & Marti, 2006). Nevertheless, SE remains a promising approach for addressing social issues and promoting sustainable development globally. To fulfill its potential, SE requires ongoing support and collaboration among stakeholders, including the government, private sector, civil society, and communities themselves (Nicholls & Cho, 2006).

According to Austin et al. (2006), SE has the potential to drive systemic change and address complex social issues that traditional approaches have struggled to resolve. Researchers have emphasized the importance of SE from multiple perspectives. For instance, Mair and Marti (2006) provided a platform for SE to challenge and redefine concepts and assumptions in other areas of management and economic research, making SE a crucial research area. Nagler (2007) praised SE for its contributions to economic development policy, especially its capacity to increase socio-economic value through job creation, innovation, and social capital provision. SE is also commended as a means to provide social goods and services to society, particularly when government agencies fail to provide essential resources and services such as employment, healthcare, and education (Peredo & McLean, 2006).

The 2006 Diamond Report revealed that social entrepreneurs comprised 3.2% of the workforce, highlighting the global significance of social enterprises (Harding & Cowling, 2006). Tan (2022) notes that social entrepreneurs are driven by a passion for making a difference through social change (Mair & Marti, 2019), unbound by traditional problem-solving approaches, fostering trust, cooperation, and collaboration for a more significant impact (Weerawardena et al., 2021). They have developed innovative and effective solutions to social issues and can promote gender equality and women's empowerment by challenging traditional gender roles and creating opportunities for women (Hunter-Jones et al., 2022). SE plays a crucial role in promoting sustainable development by focusing on long-term solutions that are financially sustainable, environmentally responsible, and socially inclusive (Radley et al., 2021). SE encourages civic engagement and democratic participation, allowing participants to contribute to democratic processes and build more robust and resilient democratic institutions while also contributing to economic growth and development by creating new markets, generating employment opportunities, and promoting innovation (Zahra et al., 2022; Zahra & Wright, 2011).

2.3 Tourism Social Entrepreneurship (TSE)

Although the primary goal of tourism businesses is profit maximization (Sheldon et al., 2017), evolving global goals, customer demands, and stakeholder pressures have led tourism businesses to commit to meeting their social responsibilities (Sharifi-Tehrani et al., 2022; Zaremohzzabieh et al., 2019). Tourism social entrepreneurship (TSE) has been recognized as a significant contributor to economic development in many countries (Mair & Marti, 2006; Sobaih, 2020). Studies have assessed the economic contribution of the TSE, highlighting its potential to create jobs, generate income, and stimulate economic growth in local communities (Hall et al., 2012). Moreover, the TSE has been found to promote inclusive economic growth by empowering marginalized communities and promoting local ownership of tourism development (Hansen et al., 2019). TSE often involve local community members in designing, implementing, and managing tourism projects, giving them a sense of ownership and control over their economic futures (Woods, 2010).

The TSE also contributes to the preservation of cultural heritage and environmental sustainability in rural areas. TSE initiatives have been linked to the preservation of the traditional knowledge and practices of indigenous communities (Budovska et al., 2020). In addition, TSE ventures have been associated with enhancing environmental sustainability by promoting responsible tourism practices (Wu et al., 2021). However, challenges exist in evaluating the socioeconomic contribution of TSE, including a lack of systematic evaluation, standardized indicators, and long-term sustainability assessments (Gössling et al., 2012).

TSE can promote sustainable practices, such as responsible waste management, natural resource conservation, and cultural heritage preservation (Dellaert, 2019). However, there are challenges in accessing funding, competition with established businesses, and policy environments that may not consider the needs of local communities (Li et al. 2020a, b). Policies that support sustainable TSE and consider the social, cultural, and environmental contexts of destinations are essential.

Social entrepreneurs in South Africa work in marginalized communities, developing innovative solutions to address specific needs, promoting social cohesion, and reducing social inequalities (Makhaye & Subban, 2021). In South Africa, SE also plays a critical role in job creation and economic development, with social enterprises creating over 200,000 jobs and contributing to local supply chains and value-added products (Ramabodu et al., 2023), and contributing toward addressing climate change, biodiversity loss, and environmental issues, promoting environmental sustainability, and contributing to sustainable development goals (Telbisz et al., 2023). However, the SE in South Africa faces challenges such as limited access to finance, regulatory barriers, and a lack of supportive infrastructure (Vilakazi et al., 2021). Addressing these challenges requires collaboration among stakeholders, including the government, private sector, and civil society (Makhaye & Subban, 2021). Gender equality in SE remains a challenge in South Africa, necessitating targeted support for women social entrepreneurs (Ramabodu et al., 2023).

2.4 Research Methodology

This study combined both qualitative and quantitative methodologies. This approach was chosen to triangulate the findings, enhance validity, and provide complementary insights into the complexities of social entrepreneurship and government support. By integrating qualitative and quantitative data, this study aims to capture both numerical patterns and contextual nuances, ultimately offering a holistic understanding of the phenomena under investigation. It addressed the "why" and "how" questions through qualitative methods and explored the "what" and "who" aspects with quantitative methods. Following a constructivist approach, the cross-sectional design offers a snapshot of this research. In the context of social entrepreneurship and government support, a constructivist approach is particularly relevant because it acknowledges the diverse perspectives and interpretations of stakeholders involved in these processes. The study's focus on Bergville, oKhahlamba Municipality, involves a census approach due to the limited number of tourism establishments encompassing all 35 registered entities. The population consists of tourism businesses and government representatives associated with them.

Data collection involved self-administered questionnaires distributed via email and physical means, spanning two months. Qualitative data were gathered through semi-structured interviews with one representative of the government and captured and thematically analyzed. Quantitative data were processed using SPSS version 26, which provides graphical representations and inferential statistics. Descriptive statistics condense information, and inferential statistics, such as correlation analysis, identify significant relationships between variables. This research methodology, combining qualitative and quantitative approaches, aimed to enhance the depth and validity of the findings.

2.5 Results and Discussion

2.5.1 Perceived Challenges of Social Entrepreneurship

This question addresses the perceived challenges faced by social entrepreneurs. This was an important question because it directly spoke to one of the research objectives. Furthermore, the literature review suggests that social entrepreneurship faces challenges. For instance, Dacin et al. (2010) highlighted the shortage of resources as one of the factors hindering SE, whereas Hockerts (2017a, 2017b) and Mair and Marti (2006) pinpointed financial and human resources as hindrances. Table 2.1 illustrates the scoring patterns for the different types of challenges they experienced. The alternatives provided were a lack of resources, support from the government, community support, and an understanding of social entrepreneurship. The scoring patterns showed that the participants responded positively to three of these challenges and negatively to one. Bacq and Janssen (2011) and Mair and Marti (2006) also allude to challenges related to government policies, funding schemes, tax incentives, and other legal frameworks that could either promote or suppress SE.

2.5.2 Challenges and Constraints in Bergville's Tourism Social Entrepreneurship

This question addressed challenges specific to SE in the study area. Variables are listed in Table 2.1, which summarizes the scoring patterns for each.

All of the chi-square p-values were significant, ranging from $p < 0.001$ to $p = 0.002$. All statements showed significantly higher levels of agreement, with a lack of understanding of SE (C2.2) and a lack of collaboration between entrepreneurs (C2.3) with higher levels of neutral scores. This means that participants felt that most of them had a negative impact on SE in the area. Dacin et al. (2010); and Short et al. (2010) opined that collaborative networks provide access to resources, expertise, and market opportunities, thereby contributing to the sustainability and scalability of social ventures.

The vast majority (74.3%) agreed that there were limited networking opportunities and opportunities to share good practices. Another 68.6% agreed that there was a lack of coordination and communication between the social entrepreneurs. These findings were supported by Dees and Anderson (2003) and Mair and Marti (2006), who stated that access to networks and learning communities facilitates the sharing of best practices, lessons learned, and support among social entrepreneurs. Lack of infrastructure and lack of skilled labor both received a 60.0% agreement score. Dacin et al. (2010); Zahra et al. (2022) concurred that limited infrastructure can impede the implementation and growth of social ventures, and Mair and Marti (2006); Zahra et al. (2022) who enumerated financial resources as a significant obstacle to SE.

The following is an excerpt from the qualitative data collected (description of the regulatory or legal challenges encountered while running tourism social entrepreneurship initiatives in the study area):

Participants: *Regulatory and legal challenges can arise when running tourism social entrepreneurship initiatives. Some common challenges include (a) Ambiguity or lack of specific regulations governing social enterprises in the tourism*

Table 2.1 Challenges of tourism social entrepreneurship in Bergville

		Strongly disagree		Disagree		Neutral		Agree		Strongly agree		
		Count	Row N (%)	Count	Row N (%)	Count	Row N (%)	Count	Row N (%)	Count	Row N (%)	Chi-square p-value
Funding is a challenge to the successful implementation of SE	C2.1	0	0.0	1	2.9	13	37.1	0	0.0	21	60.0	< 0.001
Lack of awareness and understanding of social entrepreneurship among the local community in Bergville	C2.2	0	0.0	5	14.3	22	62.9	0	0.0	8	22.9	0.001
Lack of collaboration between tourism social entrepreneurship initiatives and other local businesses in Bergville	C2.3	1	2.9	4	11.4	18	51.4	0	0.0	12	34.3	<0.001
Limited opportunities for networking and sharing of best practices among social entrepreneurs in Bergville	C2.4	0	0.0	1	2.9	0	0.0	26	74.3	8	22.9	<0.001
Lack of infrastructure and resources available to support tourism social entrepreneurship in Bergville	C2.5	0	0.0	3	8.6	0	0.0	21	60.0	11	31.4	0.001
Lack of skilled labor and workforce development in Bergville, hindering the growth of tourism social entrepreneurship	C2.6	0	0.0	4	11.4	0	0.0	21	60.0	10	28.6	0.002
Lack of coordination and communication between social entrepreneurs and government agencies in Bergville	C2.7	0	0.0	3	8.6	0	0.0	24	68.6	8	22.9	<0.001

sector. (b) Complex licensing and permitting processes that can be burdensome for small-scale social enterprises. (c) Difficulty in navigating overlapping or conflicting regulations from different government departments or agencies. (d) Limited recognition or accreditation frameworks for social enterprises, which may hinder their access to certain benefits or support.

The interviews revealed several important challenges encountered by social entrepreneurs in the tourism industry. Some of these are discussed in the literature review section. These included ambiguity or lack of specific regulations governing social enterprises in the tourism sector, which Van Hemert et al. (2020) argued should be clear and specifically address the unique characteristics and activities of social enterprises in the tourism sector; complex licensing and permitting processes, which Wang and De Lacy (2020) warned should be simplified and streamlined, particularly for small-scale social enterprises; difficulty in navigating overlapping and conflicting regulations, which Li et al. (2020a, b) felt could be addressed through interdepartmental coordination and collaboration to ensure coherence and consistency in regulations; and limited recognition or accreditation frameworks for social enterprises, which Chan et al. (2020) argued could be resolved through the development of recognized accreditation frameworks specific to social enterprises in tourism to enhance their visibility, credibility, and access to funding and resources.

2.5.3 Cross-Tabulation of SE's Potential to Improve the Economies of Bergville vs Lack of Resources (Financial, Human, etc.)

In this section (see Table 2.2), the researcher used cross tables to analyze information about the relationship between the selected variables (social entrepreneurship potential to improve the economies of Bergville * lack of resources (financial, human, etc.). Chi-square tests of independence were performed to determine whether there was a statistically significant relationship between the variables (rows vs. columns).

The table indicates that 88.9% of 8 ($n = 7$), (agree + strongly agree) did not seem to think that lack of resources (financial and human) changed their perception that SE has the potential to improve the economies of Bergville. However, another 96.1% of 25 ($n = 24$), (agree + strongly agree) seemed to think that it did. Slightly more respondents indicated that a lack of resources (financial, human, etc.) had a bearing on their belief that SE has the potential to improve Bergville's economies ($p = 0.189$).

2.5.4 Role of Government in Supporting and Facilitating TSE in the Area

This section explores the perspectives shared by the participants regarding the government's role in supporting and facilitating tourism social entrepreneurship in the area. Various viewpoints expressed by the participants, including positive contributions, mixed views, concerns, and recommendations, were analyzed. By analyzing these perspectives, the envisaged role of the government in fostering social entrepreneurship in the tourism sector and identifying potential areas for improvement is clarified. Participants' viewpoints on the role of the government are captured in Fig. 2.1.

Participants in the study emphasized the importance of government support and facilitation for tourism social entrepreneurship (TSE) initiatives within the area. This aligns with existing literature highlighting the role of the government in fostering an enabling environment for sustainable tourism development and social entrepreneurship. For instance, Morrison et al. (2013) discuss the significance of government policies and regulations in promoting social entrepreneurship in tourism, emphasizing the need for supportive frameworks to address regulatory barriers and provide financial incentives. Similarly, Mair and Marti (2006) argued that government interventions are essential for creating conditions conducive to social entrepreneurship, including access to resources, infrastructure, and networking opportunities. Hall et al. (2012) highlight the importance of stakeholder collaboration and government involvement in driving sustainable tourism initiatives, underscoring the role of partnerships between government agencies, private sector actors, and civil society organizations. These insights underscore the need for

Table 2.2 Crosstab: social entrepreneurship and implications for economies

			Lack of resources (financial, human, etc.)		
			No	Yes	Total
Social entrepreneurship has the potential to improve the economies of Bergville	Strongly disagree	Count	1	0	1
		% within lack of resources (financial, human, etc.)	11.1%	0.0%	2.9%
	Disagree	Count	0	1	1
		% within lack of resources (financial, human, etc.)	0.0%	3.8%	2.9%
	Agree	Count	5	5	10
		% within lack of resources (financial, human, etc.)	55.6%	19.2%	28.6%
	Strongly agree	Count	3	20	23
		% within lack of resources (financial, human, etc.)	33.3%	76.9%	65.7%
Total		Count	9	26	35
		% within lack of resources (financial, human, etc.)	100.0%	100.0%	100.0%

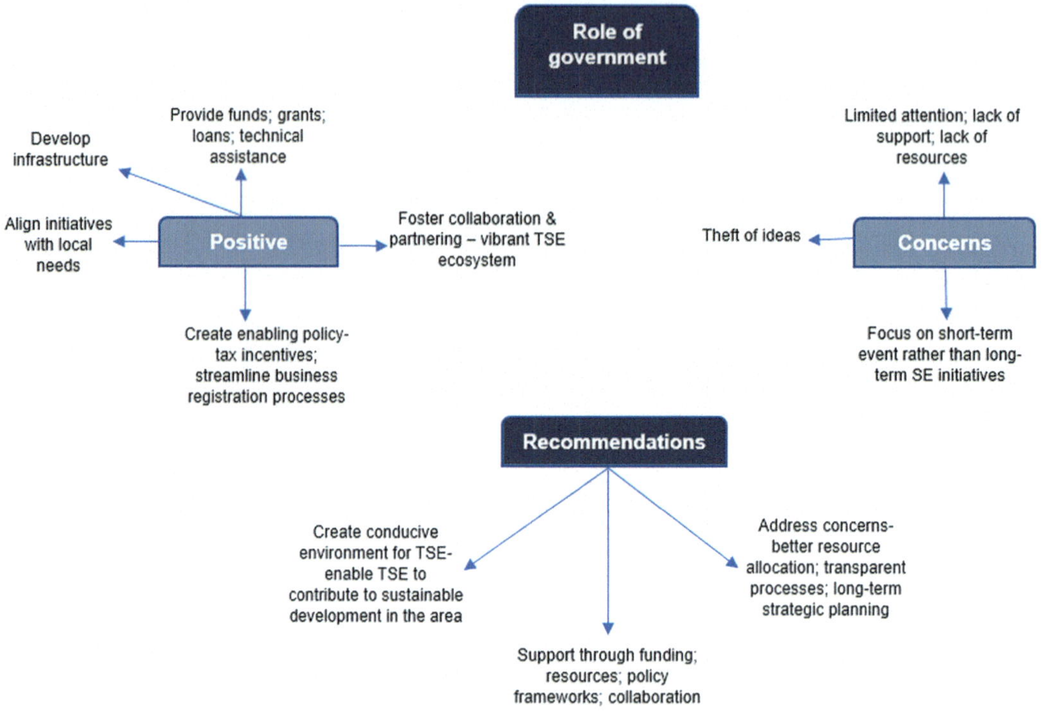

Fig. 2.1 Participants' suggestions on the government's role in supporting and facilitating TSE in the area

further research to explore the specific mechanisms through which the government can support and facilitate TSE initiatives, considering the contextual dynamics and challenges within the area.

Positive contributions from the government, such as providing funding, collaboration with the community, and infrastructure development, align with theories of government support for entrepreneurship. Studies have highlighted the importance of government funding and resources in promoting entrepreneurship (Audretsch & Keilbach, 2005). The provision of grants, loans, and technical assistance mentioned by respondents reflects the government's role in facilitating the growth and development of social entrepreneurship initiatives (Mair & Marti, 2019).

The idea of creating an enabling policy environment, including tax incentives and streamlined business registration processes, is supported by research on the role of government policies in fostering entrepreneurship (Naudé, 2018). Such policies can help reduce barriers to entry, stimulate innovation, and encourage social entrepreneurship in the tourism sector.

The significance of collaboration and partnerships, emphasized by the respondents, is aligned with the actor-network theory used to underpin the study. Collaborative approaches, facilitated by the government, can foster social innovation and create a supportive ecosystem for social entrepreneurship (Mason & Brown, 2014). By bringing together social entrepreneurs, local communities, and other stakeholders, the government can contribute to a more inclusive and sustainable tourism industry (Dacin et al., 2010).

The acknowledgment of infrastructure development as a positive contribution from the government is supported by studies of the role of infrastructure in tourism development. Infrastructure investments, such as tourism events and improvements in transportation and amenities, can attract visitors and stimulate local businesses (Ribeiro-Soriano, 2017). This aligns with theories on tourism-led local economic development, in which infrastructure development is recognized as a catalyst for tourism entrepreneurship (Ruhanen et al., 2015).

On the other hand, concerns raised by some respondents regarding the government's limited attention, lack of funding support, and inadequate support for social entrepreneurship find resonance in studies on the challenges faced by social entrepreneurs. Research has highlighted the need for improved government support, resource allocation, and transparent processes to foster social entrepreneurship (Zahra et al., 2022).

From the qualitative data collected on government initiatives and programs aimed at supporting and facilitating TSE, the excerpt below is worth noting:

Participant: *In our area, we have implemented several government initiatives and programs to support and facilitate tourism social entrepreneurship. One notable initiative is the establishment of a dedicated fund that provides financial assistance and resources to individuals and organizations engaged in tourism social entrepreneurship. Additionally, we have collaborated with local tourism boards and industry associations to develop mentorship programs, training workshops, and networking events specifically tailored for social entrepreneurs in the tourism sector.*

The initiatives undertaken in the Bergville area align with the principles outlined in the literature on the support and facilitation of tourism social entrepreneurship. These efforts encompass various forms of support, including financial assistance, provision of resources, mentorship opportunities, and networking platforms. Such initiatives indicate proactive measures by the local government to create an enabling environment for the growth and success of social tourism entrepreneurs.

Specifically, the key elements highlighted in these initiatives include the establishment of a dedicated fund tailored for Tourism Social Entrepreneurship (TSE), fostering collaboration among stakeholders, implementation of mentorship programs, organization of training workshops, and hosting networking events. These aspects, as emphasized by Dees and Anderson (2006), underscore the importance of access to capital for social entrepreneurs. Similarly, Defourny and Nyssens (2010) highlighted the significance of training and mentorship programs in enhancing entrepreneurial skills and knowledge. Furthermore, Hockerts (2017a, 2017b) emphasizes the role of networking events and collaborations with industry associations in providing social entrepreneurs with valuable connections, knowledge exchange opportunities, and access to business prospects.

2.6 Discussions

Addressing the perceived challenges in social entrepreneurship requires a multifaceted approach that, draws upon the strategies proposed in the existing literature. Social entrepreneurs often grapple with resource constraints, particularly in terms of financial and human resources. To address this, dedicated funds and grant programs established by governments or philanthropic organizations can provide much-needed financial support (Bacq & Janssen, 2011). Collaborative efforts involving governments, social entrepreneurs, and experts in the field are pivotal in creating supportive policies. These efforts can simplify bureaucratic processes, provide tax incentives, and establish a conducive legal environment for social entrepreneurship (Hockerts, 2017a, 2017b; Mair & Marti, 2006).

Raising awareness of social entrepreneurship's potential benefits and engaging the local community as stakeholders are essential to combat the challenges of community support and understanding. This involves community education, workshops, and outreach programs. In addition, initiatives to build the skills and capacity of social entrepreneurs are essential. This can be realized through training programs, mentorship initiatives, and networking opportunities (Smith et al., 2021).

Inadequate infrastructure can be a limiting factor affecting the reach and impact of social enterprises. To address this issue, government agencies should invest in infrastructure development, including transportation, communication, and utilities, to improve the accessibility of tourism destinations and foster economic growth (Allison et al., 2022). Effective collaboration among different stakeholders, including governments, communities, and industry associations, is often lacking but necessary. Fostering partnerships and multi-stakeholder collaborations can create an ecosystem that supports social entrepreneurship. This includes sharing resources, knowledge, and decision-making (Emerson et al., 2012).

To address the funding challenge in Bergville's tourism social entrepreneurship (C2.1), local social entrepreneurs can explore various sources of funding, including grants, loans, and technical assistance provided by government initiatives (Mair & Marti, 2019). They can also seek external funding through impact investors, crowdfunding, or partnerships with socially responsible businesses (Dacin et al., 2010). These diverse funding strategies can help overcome the financial obstacles faced by social ventures.

Promoting awareness and understanding of social entrepreneurship (C2.2) can be achieved through collaborative effort. Social entrepreneurs can work with local educational institutions and community organizations to promote awareness and understanding of social entrepreneurship in the community (Short et al., 2010). Public awareness campaigns, workshops, and educational initiatives can be organized to educate the local population about the benefits and potential of social entrepreneurship and foster a supportive environment for these initiatives (Short et al., 2010).

Collaboration is essential for the success of social entrepreneurship (C2.3). Social entrepreneurs should actively collaborate with local businesses in Bergville (Dacin et al., 2010). Government agencies can play a pivotal role in facilitating partnerships by connecting social entrepreneurs with local businesses and fostering a collaborative culture of collaboration (Dacin et al., 2010; Mason & Brown, 2014). These collaborations can lead to shared resources, expertise, and market opportunities, thereby contributing to the sustainability and scalability of social ventures.

Enhancing networking opportunities (C2.4) can be achieved through the establishment of initiatives and organizations that provide platforms for social entrepreneurs to network and share best practices (Mair & Marti, 2019). Government-supported networking events can be organized to connect social entrepreneurs with potential partners, mentors, and investors (Hockerts, 2017a, 2017b). Networking opportunities are crucial for knowledge exchange and business growth in the tourism sector.

Infrastructure development (C2.5) can result from the collaboration between the government and social entrepreneurs. Public-private partnerships can be explored to secure the necessary resources for infrastructure projects in the tourism sector (Page et al., 2014). Collaboration can lead to the development of tourism-related infrastructure, attracting visitors and stimulating local businesses, thereby benefiting the tourism social entrepreneurship ecosystem.

Workforce development (C2.6) can be addressed by local educational institutions offering training programs to develop a skilled workforce for social entrepreneurship in Bergville (Dacin et al., 2010). Social entrepreneurs can collaborate with these institutions to provide hands-on training opportunities (Defourny & Nyssens, 2010). These programs can enhance individuals' entrepreneurial skills and knowledge, enabling them to contribute effectively to the tourism social entrepreneurship sector.

The coordination and communication between social entrepreneurs and government agencies (C2.7) can be improved by establishing clear communication channels and platforms for regular interactions (Mair & Marti, 2019). Social entrepreneurs should actively participate in dialogue with government agencies and advocate for improved coordination and support (Mair & Marti, 2019). Effective coordination is essential to address regulatory challenges and ensure the success of social ventures.

To address regulatory challenges, social entrepreneurs should engage with local policymakers and advocate clear and specific regulations governing social enterprises in the tourism sector (Van Hemert et al., 2020). Collaboration between social entrepreneurs and government agencies is essential to simplify and streamline the licensing and permitting processes (Wang & De Lacy, 2020). These steps can lead to a more conducive regulatory environment for tourism and social entrepreneurship initiatives.

Government initiatives, such as dedicated funds, can be strengthened and expanded to provide increased financial support for social entrepreneurs (Dees & Anderson, 2006). Additionally, mentorship programs, training workshops, and networking events should be regularly updated and improved to meet the evolving needs of social entrepreneurs (Hockerts, 2017a, 2017b). By actively engaging with these initiatives and collaborating with government bodies, social entrepreneurs can access vital resources and support to foster their initiatives and contribute to the sustainable development of Bergville's tourism sector. These actions align with established recommendations and best practices in social entrepreneurship.

2.7 Conclusion

This study constitutes a novel and profound contribution to the realm of sustainable tourism by illuminating the potent and transformative capacity of tourism social entrepreneurship to catalyze community development. In light of key dimensions and the provision of actionable recommendations, this research provides invaluable insights and guidance to an array of stakeholders, including policymakers, practitioners, and community members. By embracing and applying the findings and recommendations of this study, these stakeholders can promote substantive community involvement, safeguard cultural heritage, elevate community well-being, nurture robust partnerships, and adopt a stance of continuous learning and adaptive management. In the extensive framework of these collective endeavors, sustainable community development is driven by the implementation of tourism social entrepreneurship initiatives, thereby engendering a constellation of positive economic, social, and environmental outcomes. Future research on tourism social entrepreneurship and sustainable community development could explore various avenues to advance our understanding of the field. These include conducting long-term impact assessments to gauge the sustained effects of tourism social entrepreneurship on community development indicators, such as economic growth, social well-being, and environmental sustainability. Additionally, examining stakeholder collaboration dynamics, innovation and technology adoption, policy and regulatory frameworks, cross-cultural perspectives, scalability and replicability of successful initiatives, resilience and adaptive capacity of communities, and ethical considerations within tourism social entrepreneurship ventures are promising areas for future investigation. Delving into these aspects can provide valuable insights for policymakers, practitioners, and researchers seeking to promote sustainable tourism practices and to foster inclusive community development.

References

Allison, T. H., Anglin, A. H., Davis, B. C., Oo, P., Seyb, S. K., Short, J. C., & Wolfe, M. T. (2022). Standing out in a crowd of victim entrepreneurs: How entrepreneurs' language-based cues of personality traits affect public support. *Journal of Small Business Management, 62*(2), 1–40.

Aquino, R. S. (2022). Community change through tourism social entrepreneurship. *Annals of Tourism Research, 95*, 103442.

Audretsch, D. B., & Keilbach, M. (2005). Entrepreneurship capital and regional growth. *The Annals of Regional Science, 39*, 457–469.

Austin, J., Stevenson, H., & Wei-Skillern, J. (2006). Social and commercial entrepreneurship: Same, different, or both? *Entrepreneurship Theory and Practice, 30*(1), 1–22.

Bacq, S., & Janssen, F. (2011). The multiple faces of social entrepreneurship: A review of definitional issues based on geographical and thematic criteria. *Entrepreneurship & Regional Development, 23*(5-6), 373–403.

Banks, J. A. (1987). The social studies, ethnic diversity, and social change. *The Elementary School Journal, 87*(5), 531–543.

Budovska, V., Torres Delgado, A., & Øgaard, T. (2020). Pro-environmental behaviour of hotel guests: Application of the theory of planned behaviour and social norms to towel reuse. *Tourism and Hospitality Research, 20*(1), 105–116.

Celebi, D., Pirnar, I., & Eris, E. D. (2020). Bibliometric analysis of social entrepreneurship in gastronomy tourism. *Tourism: An International Interdisciplinary Journal, 68*(1), 58–67.

Chan, J. W., Ho, T. S., & Lo, W. K. (2020). Building accreditation frameworks for social enterprises in the tourism sector: A case study of Hong Kong. *Journal of Sustainable Tourism, 28*(4), 473–493.

Crenshaw, K. (1989). Demarginalizing the intersection of race and sex: A black feminist critique of antidiscrimination doctrine, feminist theory and antiracist politics. *University of Chicago Legal Forum, 1989*(1), 139–167.

Dacin, P. A., Dacin, M. T., & Matear, M. (2010). Social entrepreneurship: Why we don't need a new theory and how we move forward from here. *Academy of Management Perspectives, 24*(3), 37–57.

Dahlin, K. B., Chuang, Y. T., & Roulet, T. J. (2018). Opportunity, motivation, and ability to learn from failures and errors: Review, synthesis, and ways to move forward. *Academy of Management Annals, 12*(1), 252–277.

De Lange, D., & Dodds, R. (2017). Increasing sustainable tourism through social entrepreneurship. *International Journal of Contemporary Hospitality Management, 29*(7), 1977–2002.

Dees, S. (2001). The opening policy in China: Simulations of a macroeconometric model. *Journal of Policy Modeling, 23*(4), 397–410.

Defourny, J., & Nyssens, M. (2010). Conceptions of social enterprise and social entrepreneurship in Europe and the United States: Convergences and divergences. *Journal of Social Entrepreneurship, 1*(1), 32–53.

Della Porta, D., & Diani, M. (2019). *The Oxford handbook of social movements*. Oxford University Press.

Dellaert, B. G. (2019). The consumer production journey: Marketing to consumers as co-producers in the sharing economy. *Journal of the Academy of Marketing Science, 47*(2), 238–254.

Dorado, S. (2006). Social entrepreneurial ventures: Different values so different process of creation, no? *Journal of Developmental Entrepreneurship, 11*(4), 319–343.

Drucker, P. (2001). The next society. *The Economist, 1*, 1–4.

Gössling, S., Scott, D., Hall, C. M., Ceron, J. P., & Dubois, G. (2012). Consumer behaviour and demand response of tourists to climate change. *Annals of Tourism Research, 39*(1), 36–58.

Hall, C. M., Prayag, G., & Amore, A. (2012). Pro-environmental tourist behavior in the context of climate change: An analysis of inter-relationships among key determinants. *Tourism Management, 33*(1), 165–175.

Hansen, M., Hjalager, A. M., & Fyall, A. (2019). Adventure tourism innovation: Benefitting or hampering operations? *Journal of Outdoor Recreation and Tourism, 28*, 100253.

Harding, R., & Cowling, M. (2006). Points of view assessing the scale of the equity gap. *Journal of Small Business and Enterprise Development, 13*(1), 115–132.

Hockerts, K. (2017a). Entrepreneurship, development, and the spatial and social organization of work. In *Social entrepreneurship* (pp. 77–107). Palgrave Macmillan.

Hockerts, K. (2017b). Determinants of social entrepreneurial intentions. *Entrepreneurship Theory and Practice, 41*(1), 105–130.

Hockerts, K., Hehenberger, L., Schaltegger, S., & Farber, V. (2022). Defining and conceptualizing impact investing: Attractive nuisance or catalyst? *Journal of Business Ethics, 179*(4), 937–950.

Hunter-Jones, P., Sudbury-Riley, L., & Al-Abdin, A. (2022). Understanding the relationship between terminal illness and tourism: An exploratory study. *Tourism Management, 88*, 104397.

Jasper, J. M. (2014). *Protest: A cultural introduction to social movements*. John Wiley & Sons.

Kanellou, A., & Spathis, C. (2013). Accounting benefits and satisfaction in an ERP environment. *International Journal of Accounting Information Systems, 14*(3), 209–234.

Khan, S. A. R., Yu, Z., Belhadi, A., & Mardani, A. (2020). Investigating the effects of renewable energy on international trade and environmental quality. *Journal of Environmental Management, 272*, 111089.

Kim, D., Lee, J. Y., Yang, J. S., Kim, J. W., Kim, V. N., & Chang, H. (2020). The architecture of SARS-CoV-2 transcriptome. *Cell, 181*(4), 914–921.

Kirzner, I. M. (1999). Creativity and/or alertness: A reconsideration of the Schumpeterian entrepreneur. *The Review of Austrian Economics, 11*(1-2), 5–17.

Li, C., Mirosa, M., & Bremer, P. (2020a). Review of online food delivery platforms and their impacts on sustainability. *Sustainability, 12*(14), 5528.

Li, W., Wu, X., & Li, C. (2020b). Regulatory hurdles for social enterprises in China: Overcoming challenges through interdepartmental coordination. *Public Management Review, 22*(8), 1163–1184.

Littlewood, D., & Holt, D. (2018). Social entrepreneurship in South Africa: Exploring the influence of environment. *Business & Society, 57*(3), 525–561.

Snihur, Y., Thomas, L. D., & Burgelman, R. A. (2018). An ecosystem-level process model of business model disruption: The disruptor's gambit. *Journal of Management Studies, 55*(7), 1278–1316.

Mair, J., & Marti, I. (2006). Social entrepreneurship research: A source of explanation, prediction, and delight. *Journal of World Business, 41*(1), 36–44.

Mair, J., & Marti, I. (2019). Social entrepreneurship research: A source of explanation, prediction, and delight. *Journal of World Business, 54*(6), 706–712.

Makhaye, A. C., & Subban, M. (2021). Entrepreneurship and business development in the 'new normal': Can small businesses adapt? In Social sciences international research conference (p. 649).

Mason, C., & Brown, R. (2014). Entrepreneurial ecosystems and growth oriented entrepreneurship.

Mody, M., & Day, J. (2014). Rationality of social entrepreneurs in tourism: Max Weber and the sociology of tourism development. *International Journal of Tourism Anthropology, 3*(3), 227–244.

Morrison, A., Rimmington, M., & Williams, C. (2013). Social entrepreneurship in tourism: An exploratory study of UK tourism social enterprises. *Tourism Management, 36*, 13–24.

Nagler, J. (2007). *The importance of social entrepreneurship for economic development policies*. University of New South Wales.

Naudé, W. (2018). Entrepreneurship and economic development: Theory, evidence and policy. *Journal of Business Venturing Insights, 9*, 55–66.

Page, S. J., Getz, D., & Hall, C. M. (2014). *The Routledge handbook of transport economics*. Routledge.

Peredo, A. M., & McLean, M. (2006). Social entrepreneurship: A critical review of the concept. *Journal of World Business, 41*(1), 56–65.

Portales, L., & Portales, L. (2019). Business models of social enterprises. In *Social innovation and social entrepreneurship: Fundamentals, concepts, and tools* (pp. 97–114). Springer.

Radic, A., Arjona-Fuentes, J. M., Ariza-Montes, A., Han, H., & Law, R. (2020). Job demands–job resources (JD-R) model, work engagement, and well-being of cruise ship employees. *International Journal of Hospitality Management, 88*, 102518.

Radley, P. M., Davis, R. A., & Doherty, T. S. (2021). Impacts of invasive rats and tourism on a threatened island bird: the Palau Micronesian Scrubfowl. *Bird Conservation International, 31*(2), 206–218.

Ramabodu, B., Mashau, P., & Farisani, T. R. (2023). The drivers of innovation at universities: A case of South African universities. *International Journal of Professional Business Review, 8*(8), 1392.

Ribeiro-Soriano, D. (2017). Small business and entrepreneurship: Their role in economic and social development. *Entrepreneurship & Regional Development, 29*(1-2), 1–3.

Ruhanen, L., Weiler, B., Moyle, B. D., & McLennan, C. L. J. (2015). Trends and patterns in sustainable tourism research: A 25-year bibliometric analysis. *Journal of Sustainable Tourism, 23*(4), 517–535.

Saiz-Alvarez, J. M., Palma-Ruiz, J. M., Valles-Baca, H. G., & Fierro-Ramírez, L. A. (2021). Knowledge management in the esports industry: Sustainability, continuity, and achievement of competitive results. *Sustainability, 13*(19), 10890.

Santos, F. M. (2012). A positive theory of social entrepreneurship. *Journal of Business Ethics, 111*(3), 335–351.

Sharifi-Tehrani, M., Seyfi, S., & Zaman, M. (2022). At the intersection of tourism social entrepreneurship and empathy: Development and validation of an empathy scale. *Journal of Business Research, 141*, 433–447.

Short, J. C., Moss, T. W., & Lumpkin, G. T. (2010). Research in social entrepreneurship: Past contributions and future opportunities. *Strategic Entrepreneurship Journal, 4*(1), 85–99.

Smith, B. R., McMullen, J. S., & Cardon, M. S. (2021). Toward a theological turn in entrepreneurship: How religion could enable transformative research in our field. *Journal of Business Venturing, 36*(5), 106139.

Snow, D. A., Soule, S. A., & Kriesi, H. (2014). *The Wiley-Blackwell companion to social movements*. John Wiley & Sons.

Telbisz, T., Imecs, Z., Máthé, A., & Mari, L. (2023). Empirical investigation of the motivation and perception of tourists visiting the Apuseni nature park (Romania) and the relationship of tourism and natural resources. *Sustainability, 15*(5), 4181.

Van Hemert, M. E., Nijkamp, P., & Van den Berg, L. (2020). Ambiguity and regulatory adaptations in social entrepreneurship. *Nonprofit and Voluntary Sector Quarterly, 49*(1), 159–176.

Vilakazi, M., Burger, C. R., Mboweni, L., Mamushiane, L., & Lysko, A. A. (2021). Evaluating an evolving OAI testbed: Overview of options, building tips, and current performance. In *2021 7th international conference on advanced computing and communication systems (ICACCS)* (Vol. 1, pp. 818–827). IEEE.

Wang, D., & De Lacy, T. (2020). The regulatory environment of small-scale social enterprises: Challenges, needs, and ways forward. *Journal of Sustainable Tourism, 28*(7), 979–995.

Woods, M. (2010). *Rural*. Routledge.

Wu, M. Y., Tong, Y., Wall, G., & Ying, T. (2021). Cultural production and transmission in museums: A social practice perspective. *Annals of Tourism Research, 87*, 103130.

Zahra, S. A., & Wright, M. (2011). Entrepreneurship's next act. *Academy of Management Perspectives, 25*(4), 67–83.

Zahra, S. A., Petricevic, O., & Luo, Y. (2022). Toward an action-based view of dynamic capabilities for international business. *Journal of International Business Studies, 53*(4), 583–600.

Zaremohzzabieh, Z., Ahrari, S., Krauss, S. E., Samah, A. A., Meng, L. K., & Ariffin, Z. (2019). Predicting social entrepreneurial intention: A meta-analytic path analysis based on the theory of planned behavior. *Journal of Business Research, 96*, 264–276.

3 Gender Mainstreaming and Dynamics of Women Participation in Tourism in African Rural Conservation Landscapes

Chiedza N. Mutanga, Joseph E. Mbaiwa, Reniko Gondo, and Olga L. Kupika

Abstract

While the literature on gender and tourism in Africa has explored aspects of gender mainstreaming and women's participation, some gaps remain in comprehensive assessments of tangible outcomes and the sustainability of gender mainstreaming initiatives. Drawing on gender mainstreaming and empowerment theories, this chapter discusses the trends and challenges associated with gender mainstreaming and women's participation in African rural conservation landscapes. This study utilized an inductive qualitative approach to extract deeper insights from an open-ended examination of secondary data sources on gender mainstreaming and women's experiences in tourism. The major trends identified included an increase in policy integration and gender mainstreaming, a growth in women's participation in tourism, an increasing trend in women-led enterprises, increasing financial empowerment among women, limited women's rights, and the persistence of gender-based inequalities. Challenges include a lack of clear implementation of policies, limited access to credit, lack of training opportunities, and cultural norms and stereotypes. To create more opportunities for women, this chapter recommends expanding access to training and capacity-building programs, diversifying tourism offerings in rural conservation landscapes, leveraging technology and online platforms, improving access to financing and micro-loans, and continuously lobbying for policies that address gender inequalities, foster women's entrepreneurship, and promote equal access to resources.

Keywords

Empowerment · Gender equity · Gender mainstreaming · Sustainable development goals · Tourism development · Women's participation

3.1 Introduction

Across most regions of Africa, women continue to face economic, cultural, social, and political disadvantages (Alahira, 2014; Nutsugbodo & Mensah, 2020; Woyo & Venganai, 2022). While economic prospects have improved in some African countries, gender disparities persist and, in some cases, widen (African Development Bank, 2015a). A high level of poverty persists in many rural African regions, with women facing

C. N. Mutanga (✉) · J. E. Mbaiwa · R. Gondo · O. L. Kupika
Okavango Research Institute, University of Botswana, Maun, Botswana
e-mail: mutangac@ub.ac.bw

K. Dube et al. (eds.), *Tourism in Rural Protected Areas*, Sustainable Development Goals Series,
https://doi.org/10.1007/978-3-031-87333-1_3

discrimination in employment, economic assets, and decision-making. Women hold only one in three non-agricultural jobs, and societal norms, religious structures, and stereotypes hinder their socioeconomic progress (African Development Bank, 2015b; Onditi & Odera, 2017; Turner et al., 2015; United Nations, 2015).

In the changing global tourism landscape, the interaction between gender mainstreaming and women's participation has emerged as a crucial intersection with the transformative potential for sustainable development and equitable progress (Alarcón & Cole, 2021). Gender equality entails equal access to opportunities, outcomes, rights, and duties and obligations for all genders (UN Women, 2015). According to the African Development Bank (2015a), expanding economic opportunities for women can raise incomes, help families escape poverty, and benefit the entire economy. The Convention on the Elimination of All Forms of Discrimination Against Women urges all member states to transform societal and traditional norms while eliminating gender-based discrimination. However, discussions on democracy and human rights in African nations often prioritize self-interest and power dynamics, neglecting the needs of marginalized groups, particularly women, who are disproportionately affected by inequality and oppression (Alahira, 2014; Onditi & Odera, 2017).

Rural communities can engage in various forms of tourism, such as ecotourism, nature-based tourism, agritourism, and adventure tourism, all of which fall under the umbrella of rural tourism. Rural tourism affects gender relations within communities due to its gendered nature in construction, presentation, and consumption, often reinforcing gender inequalities. These disparities have been addressed nationally and internationally through gender mainstreaming, which institutionalizes equality by adopting gender-sensitive practices and norms. In most African rural communities, women make up most of the population (Makandwa et al., 2023). Rural women in Africa are shifting from labor to entrepreneurial roles in tourism businesses, using local resources like traditional cuisine and arts as attractions. They are also engaging in community-based tourism (CBT) to supplement domestic income (Karasi et al., 2017; Mandipaka, 2014; Mayaka, 2015). Runyowa (2017) defined CBT as the range of tourism-related activities that provide host communities the ability to manage and benefit from engaging in tourism-related ventures in their area.

Many African rural conservation communities have been forcibly relocated to establish protected areas (PAs), and they now reside near PA boundaries (Mutanga, 2022). These communities possess valuable tourism resources, including cultural and natural assets, which are essential for various rural tourism businesses such as providing amenities like lodging and food, as well as offering wildlife and cultural tours (Sibanda, 2015). These rural conservation landscapes, known for their rich natural resources and unique cultural heritage (Keitumetse, 2016; Mutanga et al., 2023), provide an intriguing context for exploring gender-tourism relationships.

As the tourism industry has increasingly become a key driver of economic growth and cultural exchange, understanding the dynamics of women's participation is essential for crafting sustainable and inclusive pathways for development. Musavengane and Leonard (2019) argue that providing women with increased access to economic opportunities is essential for securing more inclusive growth. However, while the existing literature on gender and tourism in Africa has explored aspects of gender mainstreaming and women involvement (see, for instance, Alahira, 2014; Aleshinloye, 2023; Makandwa et al., 2023; Mandipaka, 2014; Moswete & Lacey, 2015; Nutsugbodo & Mensah, 2020; Nutsugbodo et al., 2020; Nyaruwata & Nyaruwata, 2013; Onditi & Odera, 2017; Oyendo, 2020; Rudhumbu et al., 2020), there remains a gap in comprehensive assessments of tangible outcomes and the sustainability of gender mainstreaming initiatives related to women's involvement in tourism within rural communities. This gap encompasses the representation of rural women's voices and perspectives in shaping tourism policies and practices at the local level as well as the alignment between policy intentions and the actual implementation of gender mainstreaming and tangible outcomes for women in rural communities. The gap highlights that despite some advances in gender mainstreaming and women's inclusion within African tourism, rural women remain marginalized, citing their likely experience with many forms of existing commercial practices (Aleshinloye, 2023; Oyendo, 2020). From this perspective, this chapter aims to advance theoretical debate on gender mainstreaming and women's tourism participation dynamics. In particular, this chapter analyses the patterns and issues related to gender mainstreaming and women's involvement in rural African conservation landscapes. This chapter provides an overview of these trends to highlight a growing body of scholarship on gender mainstreaming and women's participation in African rural conservation landscapes, drawing attention for other scholars, not only to deepen understanding but also to make theoretical contributions based on systematic empirical research.

3.2 Theoretical Underpinnings of the Study

This chapter is informed by gender mainstreaming and empowerment theories. On the one hand, gender mainstreaming recognizes that the roots of disproportionate advantages and disadvantages among individuals along all parts on the binary gender divide originate from societal norms, power relations, patterns of socialization, and stereotypes (Alahira, 2014; Carvalho et al., 2019; Je et al., 2022). It thus aims to achieve gender equality, promote human rights for women, and result in more inclusive policies across a wide range of policy areas that include education, health, employment, and social welfare through the incorporation of gender perspectives into all decision-making processes (Nutsugbodo et al., 2020). Gender mainstreaming in tourism is a development planning and management system aimed at ensuring that both men and women reap the benefits. It also includes addressing gender-based stereotypes and discrimination in tourism marketing, women's leadership roles and education, as well as developing high-quality tourism experiences that enable economic equality for women through empowerment or involvement (Alarcón & Cole, 2021; Boonabaana, 2014).

Empowerment theory, on the other hand, comes from sociological and psychological perspectives that help explain complex processes by which individuals or communities gain control over their lives and environments, with a particular emphasis placed on personal agency, self-efficacy, and autonomy (Turner & Maschi, 2015). It underlines the ephemeral nature of empowerment, in which systems are challenged and resources appropriated. The empowerment theory thus offers an understanding into gender mainstreaming and the involvement of women in tourism (Chandran & Mohanty, 2018). It encompasses activities related to entrepreneurship, involvement in decision-making, and gender-sensitive policy advocacy, as well as other forms of economic, social, and political empowerment (Aghazamani et al., 2020; Moreno-Gavara et al., 2019). Drawing from gender mainstreaming and empowerment theories, this chapter explores existing programs and policies put in place to address gender-based inequities and discrimination while promoting women's active participatory roles in rural tourism.

3.3 Gender Mainstreaming and Women's Participation in Tourism

Gender mainstreaming has assumed greater importance in various development sectors including tourism (Boonabaana, 2017). This is in line with the international development paradigms such as Beijing Declaration and Platform for Action (1995) and SDGs, whose objectives are to eradicate gender inequalities by mainstreaming gender along all facets of development (Onditi & Odera, 2017). Gender mainstreaming in the tourism industry involves bringing about good improvements and reframing women's roles by acknowledging their contributions that go beyond stereotypes. It aims to dismantle ingrained prejudices that keep women from gaining access to opportunities for professional advancement, leadership roles, and resources in the industry (Alarcón & Cole, 2021).

Gender disparity in the tourism industry has gained substantial attention in recent years (UN Women, 2011). The United Nations World Tourism Organization (UNWTO) and the World Bank recognize the potential of tourism to promote equality in gender and empower women (Ferguson & Alarcon, 2015). However, gender issues have been inadequately incorporated into tourism policies and practices, often lacking depth in addressing fundamental gender inequality (Tucker & Boonabaana, 2012). Ferguson (2015) stressed the need for significant policy measures to strengthen the links between tourism and gender equality. Research reveals ongoing wage disparities, a higher concentration of women in the informal tourism sector, and a significant burden of unpaid labor within family driven tourism businesses (Ferguson & Alarcon, 2015; UN Women, 2011). However, gender mainstreaming has made partial progress: some gender aspects are integrated into mainstream agendas, while others are sidelined. International and national policies aim to promote gender equality in tourism. The UNWTO emphasizes women's significance in tourism (UNWTO (World Tourism Organization), 2005), and collaborations between government agencies, non-governmental organizations (NGOs), and the private sector prioritize gender mainstreaming. In some areas, women's roles in tourism have evolved from supporting staff to leadership positions in planning, management, and innovation. Women play multifaceted roles across rural and urban settings, participating in entrepreneurship, hospitality, community-based tourism, etc. (Xu, 2018).

The nexus between women's engagement in tourism and SDG accomplishments has a substantial impact on gender equality and sustainable development in general (Onditi & Odera, 2017). Gender equality is emphasized in the UN 2030 Agenda for Sustainable Development as being necessary to achieve all 17 SDGs. Empowerment is often at the core of development goals aimed at closing the gender gap. While the Millennium Development Goal (MDG) 3 placed emphasis on "promoting gender equality and empowering women," the Sustainable Development Goal (SDG) 5 is more explicitly focused on "achieving gender equality, protect, and empower women, youth, and vulnerable individuals" (Onditi & Odera, 2017). Compared to other industries, the tourism sector offers women double the chance of becoming employment providers, providing an opportunity for them to explore entrepreneurship. Given that it increases women's economic, psychological, social, and political empowerment, the tourism industry is acknowledged as a critical sector for improving gender equality and women empowerment (Carvalho et al., 2019; The World Bank, 2018). As one of the strategies to empower women successfully, tourism can be a means for enhancing household income, self-esteem, and social status (Aghazamani et al., 2020). However, despite the considerable work that women do in driving tourism forward, and even though they outnumber men both as employees and participants, persistent challenges such as access to resources, labor disparities, and cultural norms prescribing patriarchal gender roles continue to thwart efforts for a full transformation. Sociocultural norms establish the traditional roles for women in ways that limit their mobility and framing of opportunities (Alarcón & Cole, 2021).Gender disparities in education and training continue to maintain differences in pay gaps surrounding high-skilled positions. Furthermore, social inequity is said to have a detrimental impact on common-pool natural resource co-management, particularly in land-reformed communities, where it results in the exclusion of marginalized groups like youth, women, and the poor (Musavengane & Simatele, 2017).

3.4 Methods

Using secondary data sources, this chapter followed five review steps: literature search, determination of selection criteria, initial screening, full-text review, and final article selection. Through a comprehensive examination and evaluation of the various viewpoints and complex conclusions found in secondary data sources, the inductive qualitative approach was used to glean deeper insights from the body of existing literature. The first step, the literature search, which was a scoping stage resulted in the identification of 259 data sources using keywords "gender mainstreaming," "women empowerment," "women's participation in tourism," and "challenges in women's involvement in tourism." We searched for literature from 2013 to 2023, selecting this timeframe to capture the most recent developments and trends in gender mainstreaming and empowerment within rural tourism, reflecting contemporary regional initiatives and discussions. To incorporate significant foundational information, exceptions were made for key documents published before 2013, especially those published by notable organizations such as the UNWTO. The databases used for accessing the literature were Google, Google Scholar, and Web of Science. Google Scholar offers broader coverage and free accessibility, making it convenient for quick searches and interdisciplinary research, whereas Web of Science provides high-quality content with advanced search features. From the Web of Science, we retrieved only articles unavailable on Google Scholar. Additionally, Google's breadth of coverage across all types of online content, including web pages, scholarly articles, and other resources, makes it an ideal database for accessing non-governmental reports and documents not found through the Web of Science and Google Scholar. To ensure journal quality, we searched Google Scholar for journals published by reputable academic publishers.

The second step, to determine the selection criteria, included a set of clear inclusion and exclusion criteria that guided the selection and filtration of secondary data sources. The following inclusion criteria were applied: (1) data sources needed to directly align with the study's objective, focusing on gender mainstreaming and/or women's participation in rural tourism in Africa, and (2) to align with the research's geographical scope, data sources needed to offer comprehensive information on one or more African countries, an overview of Africa as a whole, or one or more African regions. Sources that encompassed two or more continents were also considered if Africa was included among the covered continents. Conversely, some exclusion criteria were applied in tandem: (1) data sources that did not squarely address women's participation in tourism, gender mainstreaming, community involvement in tourism, or women empowerment were omitted; (2) materials lacking academic credibility, such as opinion pieces, editorials, and speculative content, were excluded; and (3) to remain consistent with the study's geographical scope, data sources that concentrated solely on regions outside Africa were deemed inapplicable and excluded from consideration. However, approximately 7 literature sources that did not specifically cover Africa but provided a theoretical background of the main concepts of gender mainstreaming, women's empowerment, and SDGs were included.

The third step involved initially screening the titles and abstracts of 259 documents to quickly eliminate irrelevant articles, focusing on those aligned with the research's emphasis on gender mainstreaming and women's involvement in tourism in Africa. A total of 163 documents were eliminated, leaving 96 relevant ones. In the fourth step, 96 documents underwent a full-text review and 31 articles that did not meet the inclusion criteria were excluded. The final step involved the selection of 65 secondary data sources, including journal articles, academic books, book chapters, theses, conference proceedings, and other online materials (Table 3.1).

Thematic analysis was used to identify and interpret patterns in qualitative information obtained from secondary sources. After a thorough analysis of the 65 data sources that were chosen, a coding procedure was started to identify the sections that were pertinent to gender mainstreaming and women's participation in rural tourism in Africa. This process resulted in various themes representing positive and negative trends as well as challenges. To ensure the credibility of findings, we sought feedback from colleagues not directly involved in the research process (Nowell et al., 2017).

Table 3.1 Data sources and distribution

Data source	Number	Database		
		Google scholar	Web of science	Google
Online source[a]				
Journal articles	38	29	9	
Academic books	1			1
Book chapters	6			6
Theses	2	2		
Conference proceedings	3	3		
Other online sources[b]	14			14
Offline source[a]				
Print book	1			

[a] ***Notes*:** An online source is a digital resource accessed via the Internet, whereas an offline source is not accessed digitally

[b] Other online sources refer to reports or other documents published online by organizations, institutions, or researchers that contain research findings or analyses

3.5 Results

3.5.1 Trends in Gender Mainstreaming and Rural Women's Participation in Tourism in Africa

Across Africa, laws and rights-based approaches designed to promote Gender Equality have been embedded within a number of legal frameworks that meet the requirements of United Nations SDG 5. This commitment has led to the development of gender-responsive policies in the tourism sector. For example, Lesotho implemented legal frameworks and ratified international conventions on gender mainstreaming to enhance gender relations (Kali, 2015). Women's empowerment initiatives are crucial for addressing rural poverty in Africa (Okemakinde, 2014). The Agenda 2063 by the African Union recognizes tourism as a stimulus for socioeconomic enhancement and inclusion, emphasizing the involvement of women and youth (UNCTAD, 2015). Several African governments have gradually improved the economic independence and political empowerment of women. Kenya's Vision 2030, for instance, advocates for gender mainstreaming in government budgets, programs, and policies (Oyendo, 2020). In Cameroon, legislation ensuring equal opportunities for men and women in enterprise ownership has provided reassurance to women in unskilled labor (Kimbu & Ngoasong, 2016). In the Southern African Development Community (SADC) region, South Africa, Botswana, Namibia, and Zimbabwe have implemented national gender policies to infuse gender considerations into all levels of society and the economy (Nyaruwata & Nyaruwata, 2013).

Africa has also witnessed a growth in the participation of women in tourism. In the tourism sector in Ghana, women constitute 65% of the workforce, significantly contributing to entrepreneurial opportunities for women in the country (Ghana Tourist Board, 2008). Globally, Ethiopia and Botswana rank among the countries where women manage over half of the businesses in the hospitality sector (UN Women, 2011). In African rural conservation areas, women engage in various CBT ventures and benefit from them. Notable examples include the Sengwe community in Zimbabwe's Masvingo Province and the Makuleke community in South Africa's Limpopo Province, both of which are involved in cultural tourism enterprises (Makandwa et al., 2023). Kisumu County in Kenya features female tour guides and van drivers that facilitate tourist experiences. Several capacity-building initiatives are also being implemented in many areas where destinations are embracing the concept of including women in tourism projects, particularly in community-run businesses. Chandran and Mohanty (2018) noted that active efforts are being made in Uganda to provide women with necessary skills. In the Gumbi community of South Africa, leadership and management structures demonstrate inclusivity toward women in conservation, as exemplified by the composition of the Emvokweni Community Trust board, which, as of September 2017, consisted of three males and three females, with a woman serving as chair (Musavengane & Leonard, 2019).

Another growing trend in African tourism is the rise of women-led enterprises and entrepreneurship across various subsectors in rural areas. In many rural communities across Africa, women are actively involved in various aspects of the tourism industry, including hospitality, and cultural experiences where they provide traditional meals or cultural performances for tourists, or offer guided tours showcasing local customs and traditions. This trend reflects a shift from traditional gender roles toward more diverse and economically impactful positions, aligning with empowerment theories focused on enhancing agency and economic participation. In Botswana, cultural tourism initiatives break down gender barriers and empower women by offering leadership opportunities and contributions to their communities through cultural preservation and tourism activities (Moswete & Lacey, 2015). These initiatives enable women to engage in guiding, storytelling, and showcasing cultural heritage. Similar to this, rural women entrepreneurs in South Africa's Makuleke community and Zimbabwe's Sengwe community have successfully launched and maintained ecotourism businesses over time (Makandwa et al., 2023).

Additionally, women are working to create coping strategies that will allow them to manage both their household responsibilities and tourism activities (Boonabaana, 2014). In many African rural communities, the large number of domestic duties and unpaid family labor often limits participation in tourism. These responsibilities, such as preparing food and caregiving for the family and relatives, create time constraints and reduce mobility for other activities. However, other women have managed to balance tourism and household duties by hiring helpers or asking for aid from other family members. In addition, they fight against societal inequalities that prevent them from fully engaging in the tourism industry by being cautious to prevent suspicion and reprisal from spouses, families, and communities. For example, Mukono women in Uganda manage these difficulties in spite of men's unwillingness to help out around the house (Boonabaana, 2014).

Women are challenging traditional gender stereotypes through their participation in tourism (Boonabaana, 2014). Their opportunities to participate in tourism are nevertheless limited by local gender relations, which forces them to maneuver around these restrictions cautiously and through compromise, sacrifice, struggle, and contestation. The gradual process of sociocultural gender transformation is illustrated by the way women in Mukono, Uganda, proactively confront these restraints in order to protect their survival in society (Boonabaana, 2014). Men still have the majority say when it comes to home decisions and the division of labor, but women are steadily challenging traditional notions of what it means to be a woman or a man, such as the right to own property, opppotunities to own and run businesses, the right to travel, and respect from the community. In the Sengwe community, in Zimbabwe, and the Makuleke community in South Africa, women-operated community-based tourism ventures are often located at or near their homes or shopping centers, allowing them to manage household duties, such as food preparation, alongside their venture responsibilities (Makandwa et al., 2023). This dual capability demonstrates their success in overcoming the traditional challenges imposed by customs and norms.

Women's participation in tourism has empowered them financially, leading to greater self-sufficiency for themselves and their families (Tucker & Boonabaana, 2012). Various levels of women's engagement in tourism have been observed in rural African conservation areas. In Ghana's Kakum Conservation Area, women play a vital role in ecotourism, contributing to income generation through food provision, handicraft production, and hospitality services (Nutsugbodo et al., 2020). This income boosted confidence, self-esteem, and social respect and reduced workloads for women. Women support their husbands financially, meet their families' economic needs, and acquire personal and household items independently. Economic independence has bolstered their confidence, esteem, and community respect, leading to investments in property where some women are even buying land and building houses for their families (Boonabaana, 2014).

However, while most trends are positive, there are also noticeable negative trends in gender mainstreaming and women participation in rural Africa. These include limited women's civil rights, the persistence of gender-based inequalities and discrimination against women, and a lack of confidence and active participation among women in tourism planning and decision-making processes. Ownership and control of assets can significantly impact individuals' participation in development processes, enhancing their economic value, status, and bargaining power. However, in many African countries, women's rights to assets such as land, business ownership, inheritance, and education are limited because of entrenched patriarchal norms, resulting in economic hardships and restricted access to financial opportunities (Aleshinloye, 2023). The Kenya Women Finance Trust, by extending microcredit to only 17 participants with sufficient concrete security out of a total of 190 in 2013, demonstrated how this gender disadvantage in obtaining financing even extends to the conventional banking sector (Ouma & Rambo, 2013). Many women are thus isolated from formal financial institutions in most traditional African contexts where they are not allowed to hold immovable goods like land (Ouma & Rambo, 2013; Oyendo, 2020).

Africa has a male-dominated tourism sector where women often receive lower wages and less economic prospects. In Uganda, men primarily control tourism and hospitality businesses, while rural women, constituting 90% of the agricultural sector, have fewer opportunities (Boonabaana, 2014). Women in Uganda face challenges in household decision-making, hindering their access to paid work and ownership of resources and assets. Similarly, Aleshinloye (2023) reported that while there are some recorded successes in women empowerment, Botswana is still a male-controlled society, limiting women's access to empowering educational information. In Nigeria, government policies aimed at uplifting women economically and socially have been largely ineffective, with persistent gender disparities affecting various aspects of society, including politics. Cultural and gender roles further restrict women's entrepreneurship and financial independence, pushing them into poverty (Aleshinloye, 2023).

Rural African communities are still characterized by defined gender responsibilities, with women customarily responsible for subservient roles such as housekeeping, childcare, cooking, and fetching resources, whereas men are expected to provide and make decisions. Despite efforts to promote women empowerment, social gender traditions persist as a challenge. In Uganda's Mukono community, a significant barrier to women's involvement in tourism is the need for male permission, as men prefer women to stay at home for domestic duties. Women comply to avoid conflicts, marital issues, and concerns about violence and infidelity (Boonabaana, 2014).

Women often lack confidence in participating in tourism planning and decision-making, even when opportunities are provided. In Kisumu County, Kenya, efforts are made to include marginalized groups such as women and youth through ward administrators (Oyendo, 2020). However, despite knowing about tourist planning and development, women often avoid attending meetings due to a lack of confidence or shyness, thereby missing out on valuable opportunities discussed in these gatherings, including capacity-building seminars (Oyendo, 2020).

3.5.2 Challenges

Progress in gender mainstreaming is fraught with difficulties, especially in Africa, where many policy instruments may exist on paper, but putting them into practice remains a significant challenge. Including gender in development policies is not enough to enable women's empowerment and achieve gender equity. Limited resources and the lack of clarity in how policies should be implemented or monitored can hurt desired outcomes for gender mainstreaming in tourism (Odera & Mulusa, 2020). Bianco et al. (2017) argue that policies are adopted slowly because traditional cultural norms and power structures that encourage gender discrimination have shown resilience.

The lack of education in entrepreneurship is another challenge in the empowerment of African women. Entrepreneurship education implies freeing African women from being financially dependent on men, the government, and the society. However, due to a lack of entrepreneurship education, many female entrepreneurs encounter difficulties in effectively promoting their products and services (Rudhumbu et al., 2020). According to the UNESCO (2017), women make up the majority of African adult populations and are faced with more challenges in securing training and education in tourism and hospitality. Aleshinloye (2023) opines that the target of female education has not been achieved in most African countries because tradition and religion have continued to play a significant role in keeping them subservient. Moreover, the policies and laws geared at stemming female education are not being enforced in most of these African countries (Aleshinloye, 2023).

Women's entrepreneurship in Africa is also hindered by limited access to credit, a lack of training opportunities, and disparities in educational attainment. For instance, in Nigeria, Aleshinloye (2023) reported that one of the biggest difficulties which the tourism and hospitality industry faces is a shortfall in financial resources like microcredit opportunities. This challenge is mainly caused by the fact that most lending institutions need collateral, a situation which is exacerbated by the lack personal assets that women can use as collateral to acquire money. Additionally, research shows that a significant portion of women lack access to education, which makes it difficult for them to know about funding sources and to craft funding-worthy business plans (Rudhumbu et al., 2020).

Women's access to tourism opportunities and the successful implementation of gender mainstreaming initiatives are often restricted by traditional cultural practices and stereotypes. For instance, these traditional customs prevent women from occupying roles that they have not traditionally occupied. An example is where, due to cultural grounds, the majority of African women working in the tourism industry are viewed as immoral and unrespected members of society. In other places, women are restricted from owning or inheriting land or any asset as this is regarded as the province of men (UNCTAD, 2017). Women are caught in impoverished living conditions created by these traditional practices which slows the progression of empowerment and a balance of economic stability.

Moreover, by nature, women are expected to work to provide for their families, in which case they have to reconcile their position at home (as mothers and housekeepers) with that in the workforce (UNCTAD, 2017). On the other hand, balancing tourism work with domestic responsibilities can be particularly challenging for women, further compounding their ability to fully engage in the industry (Ihugba et al., 2014). Only a small portion of women can dedicate more time and energy to their business than they do with household tasks.

Rural women's participation in tourism is also hindered by limited decision-making power, unequal benefits, and gender-based constraints (Nutsugbodo et al., 2020; Nutsugbodo & Mensah, 2020). Traditional belief systems that reinforce the subordinate status of women toward men underpin why womanhood is debarred from leadership, self-reliance, and self-constitution (Kapinga & Montero, 2017). Additionally, cultural constraints that discourage women from taking on decision-making roles within their families contribute to the development of low self-esteem and a heightened fear of failure (Gichuki et al., 2014).

Rural women face challenges in accessing telecommunication and information communication technology (ICT) such as computers and internet connectivity which make it difficult for them to keep pace with ever-changing business information, potentially negatively impacting their ventures (Ewere et al., 2015; Motilewa et al., 2015). Another barrier is the lack of or limited support from family members, which often harms women's confidence to start or run a business (Majenga & Mashenene, 2015). Rural women also have insufficient network compounds since they typically confine their networks to well-known and reliable connections (Rudhumbu et al., 2020).

3.6 Discussion

The findings highlight optimistic improvements as well as ongoing obstacles to rural African women's involvement in tourism. This aligns with multiple SDGs set by the UN. The rise in the integration of women in tourism-related activities is a definite step toward attaining SDG 5 (Gender Equality) and facilitating the accomplishment of SDG 8 (Decent Work and Economic Growth), as women are seen crossing over to nontraditional roles, functioning as independent entities with financial independence, and engaging in entrepreneurship. At the same time, they financially benefit their families, aligning with SDG 1 (No Poverty). Nonetheless, challenges are still being faced in promoting women's rights and gender equality (SDG 5 and SDG 10 [Reduced Inequalities]), for example, gender disparities in the tourism industry, the ownership of less property by women, and the underrepresentation of women in decision-making processes. These findings emphasize the necessity of gender-specific regulations and initiatives to help women succeed in the tourism industry and achieve these SDGs.

Diversity and equality have been practiced through what many authors identify as rhetoric or slogans, but they lack actual practice (Chambers et al., 2017; Je et al., 2022). The narrative can be changed by tourism stakeholders. In order to mainstream gender at national, regional, and local levels in tourism development, governments and organizations must keep campaigning for policies aimed at reducing inequalities between the sexes while encouraging women's entrepreneurial skills alongside their access to resources. While a variety of legal instruments and rights-based measures aimed at ensuring gender equality are spelled out within the AU framework as well as in many African countries, in reality, they often fail to materialize for millions of women (African Union, 2018). African Union (2018) points out that while legislative reform is necessary, it is not sufficient to ensure gender-responsive governance in practice. In fact, other deliberate strategies like gender governance systems, including gender-responsive budgeting, are also needed to foster an enabling environment for engaging rural women in tourism. According to Je et al. (2022), gender mainstreaming involves integration of gender equality policies and principles into all areas of the tourism industry.

Mentorship programs and support systems are important to increase women's involvement in the tourism industry (Madera, 2018). Governments, financial institutions, and NGOs should provide financial support in order to give rural women more opportunities to invest in their businesses and expand their product offerings. However, financial support should be accompanied by the provision of training and skills development opportunities. These capacity-building initiatives will address traditional roles such as tour guiding, but should also recognize the increasingly important role of sustainable tourism programs and digital marketing, as well as capitalizing on other evolving niche markets including ecotourism, agritourism, or cultural tourism. In line with empowerment theories, capacitating rural women is therefore important to equip them with the necessary knowledge, skills, and financial resources required for them to make informed decisions in persuing different entrepreneurial opportunities (Turner & Maschi, 2015).

It is critical that rural women receive greater recognition and representation in the tourism industry so as to reshape cultural conceptions. Following the AU Gender Strategy (African Union, 2018), this can be accomplished by highlighting and promoting women-led enterprises through media marketing and designing of tourism experiences. According to Gandiwa et al.

(2014), the media serves as the principal medium for acquainting decision-makers, interest groups, and the general public with significant development issues. As such, the publics' and policymakers' perceptions and responses to these issues are greatly influenced by the way the media presents them. The promotion of rural women's visibility in tourism can therefore aid in combating prejudices and stereotypes against women and girls, particularly since their contributions are usually disregarded.

Women are encouraged to leverage their knowledge in the tourism industry and collaborate with other colleagues, for instance through women's tourism associations, cooperatives, mentorship programs, and a variety of online programs. Rinaldi and Salerno (2020) assert that interaction and networking significantly enhance business opportunities by providing a deeper understanding of women's roles in different contexts. Technology is a powerful factor in improving access for rural women, helping them gain meaningful voices in society (Kishnani, 2021). The Organisation for Economic Co-operation and Development (Organisation for Economic Co-operation and Development (OECD), 2018) similarly emphasizes that harmonized policy measures are instrumental in closing the digital gender gap and change established stereotypes in society.

3.7 Conclusion

This chapter analyzes the trends in gender mainstreaming and women's involvement in rural African conservation landscapes. The study findings show that trends in gender mainstreaming and empowerment through women's participation in rural African tourism are indicative of a changing landscape that emphasizes inclusivity, empowerment, as well as sustainability. Policy integration, entrepreneurship driven by women, community engagement, and empowerment are mechanisms serving the continent's commitment to leverage the potential of women for a transformative role in tourism. This acknowledgment of gender mainstreaming comes in line with empowerment theories that celebrate the empowering effects of equipping women and enabling them to take part and redefine their positions within it. Although challenges remain, such as cultural norms and resource inequalities, there is an encouraging change reflected in policy evolutions, empowerment programs, and increased women visibility in rural tourism activities. However, even though there has been some achievement in terms of realizing the importance of women empowerment, there are still considerable policy implementation challenges.

This chapter concludes that gender mainstreaming and women participation in rural tourism is critical to achieving sustainable tourism development in African rural conservation landscapes as this has a significant positive impact on various SDGs. The study recommends that policymakers engage with communities and other stakeholders, such as NGOs or the private sector, in developing and implementing gender-responsive programs. The study further highlights the importance of capacity-building initiatives in terms of knowledge and skills development, collaborations and networks, governance arrangements, and infrastructure and financial resources for rural women's empowerment in tourism.

Future studies could examine the implementation of current gender mainstreaming policies and the mechanisms to enhance rural women participation in tourism to assess their capacity for empowering women's socioeconomic outcomes. These studies will also provide empirically based insights needed in order to develop effective gender mainstreaming and tourism policies that promote rural women's participation in tourism. For an in-depth comprehension of how women experience rural tourism, there is need for research on the intersection of gender with other social factors like age, ethnicity, and socioeconomic situation. Future research might also consider emerging strategies and good practices to support women-led and other and entrepreneurial ventures within tourism by leveraging technology and digital platforms that increase the access of women to markets and resources. Additionally, longitudinal and comparative studies conducted in different places can provide insight into the nature of gender relationships over time. Finally, to make the results more inclusive and equitable, it is important to include rural women in all stages of the research process through the use of participatory design principles. This will allow all voices (discrimination, prejudice, and essentially quiet aspects) to be heard.

References

African Development Bank. (2015a). *Report on 'Where are the women: Inclusive boardrooms in Africa's top listed companies'*. Immeuble CCIA.

African Development Bank. (2015b). *Empowering African women: An agenda for action (African gender equality index 2015)*. Immeuble CCIA.

African Union. (2018). *AU strategy for gender equality & women's empowerment-2018-2028*. African Union.

Aghazamani, Y., Kerstetter, D., & Allison, P. (2020). Women's perceptions of empowerment in Ramsar, a tourism destination in northern Iran. *Women's Studies International Forum, 79*, 102340.

Alahira, H. A. (2014). A re-consideration of African perspectives of women, gender, and development. *International Journal of Gender and Women's Studies, 2*, 69–86.

Alarcón, D., & Cole, S. (2021). No sustainability for tourism without gender equality. In K. Boluk, C. Cavaliere, & F. Higgins-Desbiolles (Eds.), *Activating critical thinking to advance the sustainable development goals in tourism systems* (pp. 57–73). Routledge.

Aleshinloye, K. (2023). *Women's role in Africa's tourism industry: Overcoming gender disparities, empowerment, and entrepreneurship challenges at a UNESCO World Heritage Site* (pp. 1–18). Tourism Planning & Development.

Bianco, M., Lombe, M., & Bolis, M. (2017). Challenging gender norms and practices through women's entrepreneurship. *International Journal of Gender and Entrepreneurship, 9*(4), 338–358.

Boonabaana, B. (2014). Negotiating gender and tourism work: Women's lived experiences in Uganda. *Tourism and Hospitality Research, 14*(1-2), 27–36.

Boonabaana, B. (2017). A critical analysis of tourism, gender, and poverty reduction. In A. Spenceley & D. Meyer (Eds.), *Tourism and poverty reduction: Principles and impacts in developing countries* (pp. 187–278). Routledge.

Carvalho, I., Costa, C., Lykke, N., & Torres, A. (2019). Beyond the glass ceiling: Gendering tourism management. *Annals of Tourism Research, 75*, 79–91.

Chambers, D., Munar, A., Khoo-Lattimore, C., & Biran, A. (2017). Interrogating gender and the tourism academy through epistemological lens. *Anatolia, 28*(4), 501–513.

Chandran, A., & Mohanty, P. (2018). Poverty alleviation and women empowerment through tourism development - An explorative study of model ventures. *ATNA Journal of Tourism Studies, 13*(1), 59–74.

Ewere, A., Adu, E., & Ibrahim, S. (2015). Strategies adopted by women entrepreneurs to ensure small business success in the Nkonkobe municipality, Eastern Cape. *Journal of Economics, 6*(1), 1–7.

Ferguson, L. (2015). Tourism, the millennium development goals, and gender equality. In J. D. Wright (Ed.), *International encyclopedia of the social & behavioral sciences* (2nd ed., pp. 474–480). Elsevier.

Ferguson, L., & Alarcon, D. (2015). Gender and sustainable tourism: Reflections on theory and practice. *Journal of Sustainable Tourism, 23*(3), 401–416.

Gandiwa, E., Sprangers, S., van Bommel, S., Heitkönig, I. M., Leeuwis, C., & Prins, H. H. (2014). Spill-over effect in media framing: Representations of wildlife conservation in Zimbabwean and international media, 1989–2010. *Journal for Nature Conservation, 22*(5), 413–423.

Ghana Tourist Board. (2008). *Tourism statistical fact sheet on Ghana*. Ghana Statistical Service.

Gichuki, C., Mulu-Mutuku, M., & Kinuthia, L. (2014). Performance of women-owned enterprises accessing credit from village credit and savings associations in Kenya. *Journal of Global Entrepreneurship Research, 4*, 1–13.

Ihugba, O., Odii, A., & Njoku, A. (2014). Theoretical analysis of entrepreneurship challenges and prospects in Nigeria. *International Letters of Social and Humanistic Sciences, 5*(16), 21–34.

Je, J., Khoo, C., & Yang, E. (2022). Gender issues in tourism organisations: Insights from a two-phased pragmatic systematic literature review. *Journal of Sustainable Tourism, 30*(7), 1658–1681.

Kali, M. (2015). Women empowerment in Lesotho: Myth and/or reality. *International Journal of Scientific Research and Management, 6*(3), 52–59.

Kapinga, A., & Montero, C. (2017). Exploring the socio-cultural challenges of food processing women entrepreneurs in Iringa, Tanzania, and strategies used to tackle them. *Journal of Global Entrepreneurship Research, 7*(1), 17.

Karasi, Y., Shambare, R., & Nkondo, L. (2017). Challenges faced by rural women entrepreneurs in Vhembe District: The moderating role of gender socialisation. In The 2nd annual International Conference on Public Administration and Development Alternatives (IPADA), Gaborone, Botswana.

Keitumetse, S. (2016). *African cultural heritage conservation and management: Theory and practice*. Springer International Publishing.

Kimbu, A., & Ngoasong, M. (2016). Women as vectors of social entrepreneurship. *Annals of Tourism Research, 60*, 63–79.

Kishnani, N. (2021). Digitalization and sustainable value creation in wildlife tourism. In 6th International Conference on "Economic Growth and Sustainable Development: Emerging Trends", Mysuru, India.

Madera, J. (2018). What's in it for me? Perspective taking as an intervention for improving attitudes toward diversity management. *Cornell Hospitality Quarterly, 59*(2), 100–111.

Majenga, A., & Mashenene, R. (2015). Socio-cultural factors and financial performance among women small and medium enterprises in Tanzania. *European Journal of Business Management, 6*(32), 153–163.

Makandwa, G., de Klerk, S., & Saayman, A. (2023). Culturally based community tourism ventures in Southern Africa and rural women entrepreneurs' skills. *Current Issues in Tourism, 26*(8), 1268–1281.

Mandipaka, F. (2014). Overview of women entrepreneurs in South Africa. *Mediterranean Journal of Social Sciences, 5*(9), 127–130.

Mayaka, M. A. (2015). *The role of entrepreneurship in community-based tourism*. [Doctoral dissertation, Australia: Monash University].

Moreno-Gavara, C., Jiménez-Zarco, A., & Alabi, S. (2019). Challenges and opportunities for women entrepreneurs. In C. Moreno-Gavara & I. Jiménez-Zarco (Eds.), *Sustainable fashion: Empowering African women entrepreneurs in the fashion industry* (pp. 115–145). Palgrave Macmillan.

Moswete, N., & Lacey, G. (2015). Women cannot lead: Empowering women through cultural tourism in Botswana. *Journal of Sustainable Tourism, 23*(4), 600–617.

Motilewa, B., Onakoya, O., & Oke, A. (2015). ICT and gender-specific challenges faced by female entrepreneurs in Nigeria. *International Journal of Business and Social Science, 6*(3), 97–105.

Musavengane, R., & Leonard, L. (2019). When race and social equity matters in nature conservation in post-apartheid South Africa. *Conservation and Society, 17*(2), 135–146.

Musavengane, R., & Simatele, D. (2017). Significance of social capital in collaborative management of natural resources in Sub-Saharan African rural communities: A qualitative meta-analysis. *South African Geographical Journal, 99*(3), 267–282.

Mutanga, C. (2022). Tolerance for wildlife resources through community wildlife-based tourism: Implications for sustainability. In L. Stone, M. Stone, P. Mogomotsi, & G. Mogomotsi (Eds.), *Protected areas and tourism in Southern Africa: Conservation goals and community livelihoods* (pp. 56–69). Routledge.

Mutanga, C., Kolawole, O., Gondo, R., & Mbaiwa, J. (2023). A review and SWOC analysis of natural heritage tourism in sub-Saharan Africa. *Journal of Heritage Tourism, 19*, 49–67. https://doi.org/10.1080/1743873X.2023.2255689

Nowell, L. S., Norris, J. M., White, D. E., & Moules, N. J. (2017). Thematic analysis: Striving to meet the trustworthiness criteria. *International Journal of Qualitative Methods, 16*(1), 1609406917733847.

Nutsugbodo, R., & Mensah, C. (2020). Benefits and barriers to women's participation in ecotourism development within the Kakum Conservation Area (Ghana): Implications for community planning. *Community Development, 51*(5), 685–702.

Nutsugbodo, R., Mensah, C., Amenumey, E., & Owusu, C. (2020). Women's participation in ecotourism development within the Kakum conservation area, Ghana: Implications for community planning. *Tourism Planning & Development, 17*(6), 693–710.

Nyaruwata, S., & Nyaruwata, L. (2013). Gender equity and executive management in tourism: Challenges in the Southern African Development Community (SADC) region. *African Journal of Business Management, 7*(21), 2059–2070.

Odera, J., & Mulusa, J. (2020). SDGs, gender equality and women's empowerment: What prospects for delivery. In M. Kaltenborn, M. Krajewski, & H. Kuhn (Eds.), *Sustainable development goals and human rights* (pp. 95–118). Springer.

Okemakinde, T. (2014). Women education: Implications for national development in Nigeria. *European Journal of Globalization and Development Research, 9*(1), 553–564.

Onditi, F., & Odera, J. (2017). Gender equality as a means to women's empowerment? Consensus, challenges, and prospects for post-2015 development agenda in Africa. *African Geographical Review, 36*(2), 146–167.

Organisation for Economic Co-operation and Development (OECD). (2018). *Bridging the digital gender divide: Include, upskill, innovate*. OECD.

Ouma, C., & Rambo, C. (2013). The impact of micro-credit on women-owned small and medium enterprises: Evidence from Kenya. *Global Journal of Business Research, 7*(5), 57–69.

Oyendo, H. (2020). Assessment of public sector tourism development determinants influence on gender equality in Kenya: A case of Kisumu County (Doctoral dissertation, Africa Nazarene University).

Rinaldi, A., & Salerno, I. (2020). The tourism gender gap and its potential impact on the development of the emerging countries. *Quality & Quantity, 54*(5-6), 1465–1477.

Rudhumbu, N., Du Plessis, E., & Maphosa, C. (2020). Challenges and opportunities for women entrepreneurs in Botswana: revisiting the role of entrepreneurship education. *Journal of International Education in Business, 13*(2), 183–201.

Runyowa, D. (2017). Community-based tourism development in Victoria Falls, Kompisi Cultural Village: An entrepreneur's model. *African Journal of Hospitality, Tourism and Leisure, 6*(2), 1–7.

Sibanda, M. (2015). Realms of conservation or "wildlife liberation": A case of Sengwe in Zimbabwe. *Social Dynamics, 41*(2), 253–272.

The World Bank. (2018). *Women, business, and the law*. World Bank Publication.

Tucker, H., & Boonabaana, B. (2012). A critical analysis of tourism, gender, and poverty reduction. *Journal of Sustainable Tourism, 20*(3), 437–455.

Turner, S., Cilliers, J., & Hughes, B. (2015). *Reasonable goals for reducing poverty in Africa: Targets for the post-2015 MDGs and agenda 2063 (Africa Futures*

Paper, 13, February 2015). Institute for Security Studies.

Turner, S., & Maschi, T. (2015). Feminist and empowerment theory and social work practice. *Journal of Social Work Practice, 29*(2), 151–162.

UN Women. (2011). *Global report on women in tourism 2010*. UNWTO. http://dtxtq4w60xqpw.cloudfront.net/sites/all/files/pdf/global_report_on_women_in_tourism_2010.pdf

UN Women. (2015). *Monitoring gender equality and the empowerment of women and girls in the 2030: Agenda for sustainable development: Opportunities and challenges*. Position Paper, New York, NY.

UNCTAD. (2015). *Ensuring gender-sensitive implementation of the post-2015 development framework* (Policy brief No. 5). UNCTAD.

UNCTAD. (2017). *Economic development in Africa report 2017: Tourism for transformative and inclusive growth*. UNCTAD.

UNESCO. (2017). *Literacy and non-formal education*. http://www.unesco.org/new/en/Dakar/education/literacy/

United Nations. (2015). *Review and appraisal of the implementation of the Beijing declaration and platform for action and the outcomes of the twenty-third special session of the general assembly*. United Nations.

UNWTO (World Tourism Organization). (2005). *Gender and tourism: Women's employment and participation in tourism*. UNWTO.

Woyo, E., & Venganai, H. (2022). Situating gender and disability in tourism scholarship in Africa: An introduction to intersectional perspectives. In E. Woyo & H. Venganai (Eds.), *Gender, disability, and tourism in Africa: Intersectional perspectives* (pp. 1–29). Springer International Publishing.

Xu, H. (2018). Moving toward gender and tourism geographies studies. *Tourism Geographies, 20*(4), 721–727.

4 Pro-poor Tourism in Communities Adjacent to Hwange National Park in Zimbabwe

Ndabezinhle Dube and Ngoni C. Shereni

Abstract

For most rural communities living adjacent to protected areas in Zimbabwe, small-scale agriculture was the main source of income. However, due to low annual rainfall, frequent droughts, high-input costs, restricted access to bank loans, reductions in government subsidies for farmers, and an unfavorable macroeconomic environment, they are facing increasing levels of poverty. Policymakers view wildlife-based tourism as a broad strategy to fight poverty and improve livelihoods in rural communities. This study aimed to ascertain the role of pro-poor tourism (PPT) in community development and its potential for poverty alleviation in communities living adjacent to Hwange National Park in Zimbabwe. Semi-structured interviews were conducted with 15 key informants, including government officials, private tourism operators, non-governmental organizations (NGOs), and community leadership. Thematic analysis was used to generate meanings from the interview data. The results show that there is potential for rural livelihood diversification in protected areas by utilizing tourism opportunities to raise income levels. This study's findings point to a potential connection between tourism and poverty alleviation. In addition, wildlife in protected areas provides the foundation for successful PPT, which other attractions and supporting amenities should complement.

Keywords

Pro-poor tourism · Wildlife-based tourism · Protected areas · Nature-based tourism, Hwange National Park

4.1 Introduction

Tourism has been widely promoted as a social policy development tool in most countries in the Global South. This is because of its ability to create employment, increase social capital, promote a more equitable distribution of income in society, enhance prospects of infrastructure development, and increase household income for locals (Lenao et al., 2014; Musavengane, 2018; Tervo-Kankare et al., 2018; Truong et al., 2014; Yu et al., 2019). In developing nations, traveling to promote poverty alleviation is becoming increasingly popular as a means of boosting economic growth (Scheyvens & Hughes, 2019; Truong

N. Dube (✉)
Department of Accounting and Finance, Lupane State University, Lupane, Zimbabwe

N. C. Shereni
School of Tourism and Hospitality, College of Business and Economics, University of Johannesburg, Johannesburg, South Africa

K. Dube et al. (eds.), *Tourism in Rural Protected Areas*, Sustainable Development Goals Series,
https://doi.org/10.1007/978-3-031-87333-1_4

et al., 2014). As such, pro-poor tourism (PPT) has grown in importance, as it seeks to combine the economic contribution of tourism and the participation of the poor in the tourism economy. To this end, PPT emphasizes an increase in net benefits for the less fortunate and has the poor as the primary priority (Saarinen & Rogerson, 2014; Sharpley & Naidoo, 2010). It is critical to note that PPT is not a type of tourism but an approach that applies to any form of tourism, emphasizing the benefits to the poor (Burke, 2018). Tourism patterns, such as ecotourism, community-based tourism, nature-based tourism, community-based natural resource management practices, and sustainable tourism, are synonymous with PPT because of their potential to contribute to poverty alleviation (Yu et al., 2019; Burke, 2018).

Academic discussions on PPT started gaining prominence in the 1990s as a result of the failure of top-down governance systems and increased tourism activities in poverty-stricken areas (Bakker & Messerli, 2017; Çuhadar, 2024; Matlovičová et al., 2022; Yu et al., 2019). The U.K. Department for International Development (DFID) is credited for popularizing PPT, and since then, it has been adopted widely by a wide range of stakeholders (Ashley & Goodwin, 2007; Rogerson, 2006; Scheyvens & Momsen, 2008; Scheyvens & Hughes, 2019). The realization that tourism can play a central role in fighting poverty has accelerated the advancement of PPT (Mtapuri & Giampiccoli, 2019). In addition, tourism has always been linked to regional developments in peripheral areas, resulting in it being earmarked for poverty alleviation through PPT activities (Saarinen & Rogerson, 2014).

From a multilateral policy perspective, poverty alleviation has always been the top agenda item. For example, the Millennium development goals (MDGs) list eradicating extreme poverty as Goal number 1 (Burke, 2018), and the sustainable development goals (SDGs) have "no poverty" as the first goal (Pang et al., 2024; United Nations, 2015). Furthermore, poverty alleviation is explicitly and tacitly mentioned in the other SDGs. For example, SDG 5 on gender equality, SDG 8 on decent employment and economic growth, and SDG 10 on reducing inequality within and among countries are strongly linked to SDG 1 (Chambwe et al., 2023; Scheyvens & Hughes, 2019). Several scholars have acknowledged that tourism can play a critical role in achieving poverty alleviation targets (Njoya & Seetaram, 2018; Shereni, 2022; Siakwah et al., 2019; UNWTO, 2015). Because of these developments in global policy discourse, PPT has been adopted by numerous actors, including the government, non-governmental organizations (NGOs), local communities, and tourism companies to ensure that the poor are included and benefit from tourism (Burke, 2018).

The majority of pro-poor tourist interventions occur in rural areas and are frequently associated with activities surrounding nature-based tourism (Spenceley & Seif, 2003). In most developing countries in sub-Saharan Africa, PPT is considered a viable option for poverty alleviation because such countries have wildlife resources that have the potential to attract tourists (Bakker & Messerli, 2017). However, it is important to note that PPT is also based on other resources such as township tourism, cultural and heritage tourism, and agritourism (Yu et al., 2019). Research has shown that communities located near protected areas have benefited from PPT initiatives in the past (Dube, 2019). Protected areas present numerous challenges to local communities, such as human–wildlife conflict, depredation of livestock, destruction of crops, and loss of lives and property, which increase poverty levels (Shereni & Saarinen, 2021). Therefore, tourism provides numerous opportunities for such communities to help alleviate poverty at the community level. Tourism creates demand for locally made products, creates employment at the local level, promotes diversified livelihoods in remote rural communities, and provides a market for art and craft products (Musavengane et al., 2019). Furthermore, locals can earn income by directly and indirectly participating in tourism and tax revenue collected from tourism can fund poverty alleviation activities (Njoya & Seetaram, 2018). This shows that tourism is a potent tool for fighting poverty in resource-rich, impoverished societies.

Regardless of the benefits associated with PPT, several studies have highlighted its weaknesses. Mtapuri and Giampiccoli (2019) argue that the concept is preoccupied with the poor, leaving out important considerations, such as environmental or social-cultural sustainability issues, social equity, and how the concept can be operationalized. Moreover, poverty is defined differently in different contexts, making it difficult to measure the impact of PPT (Burke, 2018; Medina-Muñoz et al., 2016). In other instances, the benefits do not accrue to the poor, but rather to the elite within the tourism system (Dang et al., 2024; Shereni & Saarinen, 2021; Truong et al., 2014). Ashley and Goodwin (2007) argue that PPT has failed to achieve its objectives for many reasons, including its limited focus at the micro-level, lack of focus on market linkages, and the absence of documented evidence showing the impact of PPT on people's livelihoods. Nevertheless, tourism has shown greater potential for poverty alleviation than other sectors because of its characteristics such as dependence on resources largely owned by the poor, providing for wider community participation, and creating linkages between tourists and local communities (Chambwe et al., 2023; Truong et al., 2014). As such, PPT initiatives have been applied widely in the sub-Saharan context, benefiting immensely communities located in tourism resource-rich areas and resulting in improved standards of living (Butler et al., 2013; Manwa & Manwa, 2014; Truong et al., 2014).

Within the Zimbabwean context, tourism has been broadly used to uplift the livelihoods of rural communities. Initiatives such as the Communal Areas Management Programme for Indigenous Resources (CAMPFIRE), introduced for communities located near protected areas to ensure that they participate and benefit from wildlife resources, were instrumental in poverty alleviation (Child, 2003; Jamal & Dredge, 2014; Mutanga et al., 2024; Shereni & Chambwe, 2024). However, there is limited research in Zimbabwe dedicated to PPT in communities adjacent to protected areas. Several studies have focused on community perceptions of community-based tourism and Community-Based Natural Resource Management practices (CBNRM) in communities surrounding protected areas (Average, 2008; Balint & Mashinya, 2006, 2008; Chiutsi & Saarinen, 2017; Dube, 2019; Gandiwa et al., 2013; Guerbois et al., 2013; Jones, 2004; Shereni & Saarinen, 2021). Poverty alleviation is a critical global issue, especially in Africa, which is one of the regions with high poverty levels (Woyo & Musavengane, 2023). Studies of this nature are, therefore, important to policymakers and developmental agencies as they provide insights into how tourism can be effectively used as a tool for poverty alleviation. The current study used communities living adjacent to Hwange National Park in Zimbabwe as a case study to ascertain the role of PPT in community development and its potential for poverty alleviation. The following sections of this chapter look at the materials and methods, presentation and discussion of findings, and conclusions.

4.2 Materials and Methods

4.2.1 Study Area Context

Hwange National Park, covering approximately 14,000 km^2, is the largest wildlife-protected area in Zimbabwe and is located in the northwest of Zimbabwe (Segobye et al., 2022). The park is an important resource for nature-based tourism in Zimbabwe, with over 100 wildlife species attracting tourists worldwide (Marunda, 2014; Matura & Mapira, 2018). Hwange National Park has one of the largest elephant populations in Africa and has an abundance of other plain-game animals (Cameron, 2018). In addition, the park is a source of livelihood for adjacent communities, providing benefits such as employment, food, firewood, construction materials, and traditional medicines, among others (Shereni & Saarinen, 2021). The communities are predominantly rural and composed of various ethnic groups, such as Nambya, Tonga, Ndebele, Dombe, Shona, and others (Guerbois et al., 2013). CAMPFIRE is one of the initiatives introduced to help eradicate poverty and ensure that local communities benefit from wildlife resources in parks (Jones & Murphree,

2004). The program has recorded mixed results in achieving conservation and poverty alleviation objectives in adjacent communities (Shereni & Saarinen, 2021). The study area is suitable for this research for two main reasons. First, it lies in a major tourism belt for the country, with many tourism activities complemented by other attractions such as Victoria Falls, Binga, Bulawayo, and Matobo National Parks. Second, the communities surrounding Hwange National Park have benefited from PPT through several community-based tourism projects that have been instrumental in improving their standards of living.

4.2.2 Data Collection and Analysis

A qualitative research design was adopted in this study to gain insight into the role of PPT in community development. Creswell (2013) argued that qualitative research helps researchers understand how participants within a certain setting perceive and address certain issues. Therefore, in this study, we used a qualitative research approach to gain a deeper understanding of PPT in the context of Hwange National Park from the perspectives of various key informants. Fifteen participants were selected from the Zimbabwe Tourism Authority (ZTA), managers from the Department of Parks and Wildlife Authority, Hwange Rural District Council (RDC), tourism operators, and local community leadership based on their knowledge of PPT within Hwange National Park. Semi-structured interviews were conducted with the selected participants using an interview guide between September and October 2022. Semi-structured interviews were deemed suitable for this study because of their ability to obtain in-depth information from the participants and because they allowed the researchers to pose follow-up questions.

The interview guide used in the data collection process comprised questions that elicited responses on the role of PPT in community development and the challenges associated with PPT in communities adjacent to Hwange National Park. The interviews lasted between 20 and 45 min, and the proceedings were audio-recorded with the permission of the participants. The data saturation point was reached at Interviewee 15, and data collection was stopped as there was no new information emanating from the participants.

Thematic analysis was employed to analyze the data collected from the interviews to identify emerging and recurring themes. The process was performed manually without the aid of qualitative data analysis software. This involved transcribing the interview recordings verbatim, coding the data by assigning short phrases, merging similar codes, and combining related codes to create themes (Braun & Clarke, 2006). Both researchers were involved in generating codes and validating themes to ensure intercoder reliability. The researchers independently developed the codes and verified the coding framework to adopt and agree on the identification of themes.

4.3 Presentation and Discussion of Findings

This section presents the results of the interviews conducted with the selected respondents and is structured in two parts. The first section examines the contribution of PPT to community development, and the second focuses on the challenges associated with PPT in Zimbabwe. Themes that emerged from the data were presented and corroborated by direct quotes from key informants.

4.3.1 The Contribution of PPT to Zimbabwean Community Development

This study sought to determine the role of PPT in community development. It was revealed that PPT plays a significant role in uplifting living standards in a community in various ways. Four themes were generated regarding this aspect: providing benefits for the less privileged, funding income-generating projects for the community, providing social services to the locals, and environmental conservation.

Concerning the first theme, providing benefits for the less privileged, expressions like "satisfying the needs of the less fortunate," "empowerment of the less privileged," "providing educational support," and "providing opportunities to the locals" were shared by the participants. The participants revealed that recurrent droughts in Hwange have resulted in poor harvests and increasing poverty levels among subsistence farmers. Therefore, tourism activities have provided significant benefits to the poor in society, helping alleviate poverty. One respondent expressed the following "*...As you know Hwange District is beset with challenges, drought is endemic in the area and an annual rainfall of about 600mm per annum has led to poor harvest and an increase in the general poverty levels among the households. The Department of Parks and Wildlife has been working with various partners and communities to come up with programs that improve people's livelihood*" as one male respondent, a nature advocate, put it (a female participant from the Department of Parks and Wildlife).

Another participant agreed, saying that "*pro-poor tourism is one way that the industry gives back to the community by hiring locals and providing grants for fees to the most vulnerable members of society*" (Male participant a tourism operator).

With regard to the second theme on funding income-generating projects for the community, it was established that funds from tourism are also channeled toward other income-generating activities, such as beekeeping, goat keeping, and horticulture. In addition to generating income for community members, these projects also help boost people's living standards and ensure food security. Various tourism companies are involved in funding income-generating projects for community members to ensure that they are self-sufficient and to reduce poverty levels. Some respondents highlighted the following: "*As an organisation, we are involved in various community outreach projects. Some notable projects are Cattle breeding, chicken projects for schools and the community, sewing projects, beads making* ...." (Male participants from the Rural District Council).

The Department of Parks and Wildlife has worked in partnership with an NGO to establish four solar-powered gardens to support year round vegetable production, the vegetables do not just provide nutrition for families but also improve income through vegetable sales' (a male respondent from the Department of Parks and Wildlife).

On the theme of providing social services, participants credited the construction of roads, schools, and clinics to tourism activities in the area. It was also indicated that tourism operators within the area contributed funds toward the provision of social services. In addition, community-based tourism enterprises run and owned by locals generate funds channeled toward social services. One respondent expressed the following: *The proceeds from the garden are used to pay incentives to teachers and repairing school amenities'* (male respondent, Kraal Head from the local community).

Another respondent asserted that "*The role of pro-poor tourism is to ensure that social and economic advantages include employment of locals, improvement of social services such as access to education and health care*" (male respondent from the Ministry of Youth Development, Indigenization, and Empowerment).

The nature-based tourism promoted in Hwange National Park is aligned with the fourth theme of environmental conservation. It was clear from the interviews that the participants connected the PPT with lessening environmental hazards, including tree cutting, harvesting wood for charcoal, and poaching animals for food. As individuals see benefits from tourism trickling to the household level, they become motivated to conserve the same environment from which they benefit. One respondent shared the following sentiments: *Poverty alleviation mechanisms that have been put in place have managed to bear tangible benefits to an extent that there is a reduction in poaching activities [....] and selling of wood harvested from the park as members of the community can now afford to have an extra income from projects that have been implemented'* (a male participant a Village Headman).

The findings of this study are in line with previous studies suggesting that PPT is largely incorporated into larger rural development projects, including activities surrounding natural resource management, and is directly tied to environmental sustainability (Braman 2001; Saville 2001). In sub-Saharan Africa, poverty and poverty alleviation issues are major areas of concern, particularly for communities surrounding national parks (Musavengane, 2018). Regardless of having wildlife resources in their vicinity, rural communities located adjacent to national parks have remained impoverished, in contrast to developments taking place within the parks (Chambwe & Saayman, 2023). In such communities, PPT has been identified as a developmental tool for improving community livelihood (Chambwe et al., 2023). Regardless of the pursuit of tourism for its economic gains, it is better than other industries in achieving poverty reduction goals (Rogerson, 2006). PPT contributes to community development in several ways, including creating employment opportunities, sustainable environmental management, infrastructure development, improving livelihoods, and empowering local communities (Rogerson, 2006). Shereni and Saarinen (2021) further revealed that income generated from tourism activities is used to fund income-generating projects that support the livelihoods of communities located near protected areas.

4.3.2 Challenges Associated with PPT in Zimbabwe

This study established several challenges associated with PPT in Zimbabwe. The challenges are categorized into three themes: lack of funding, inadequate benefits to the poor, and human–wildlife conflicts. Most participants cited that a lack of funding led to poor implementation of PPT programs, resulting in worsening poverty levels in society. Programs such as CAMPFIRE were previously supported by developmental agencies and the government. The worsening political and economic situation in the country has resulted in donors withholding funding for such projects and dwindling government funding. One interviewee observed the following: "*Another challenge is the economic crisis that the country is experiencing which makes it difficult to implement the pro-poor tourism projects*" (A female participant from the local community)

Another respondent added the following: "One major challenge is lack of funding from the government, as tourism players we need assistance because some projects that we come up with need a lot of capital injection" (A female respondent tourism operator).

The study further revealed that there are inadequate benefits from PPT accruing to the poor. Consequently, this has led to negative coping mechanisms such as poaching and selling firewood and charcoal, which destroy the park environment. Most respondents agreed that little has been done to capacitate local communities who live in proximity to Hwange National Park. The elite capture of PPT activities by those in higher positions and politically connected individuals compounded the challenges associated with pro-poor tourism. One respondent shared the following sentiments "*Notable challenges include corruption in the sense that those intended to benefit are not involved, you will notice that most people who benefit are not the poor*" (Male Respondent from the community leadership).

Human–wildlife conflicts were noted as another challenge facing communities located near Hwange National Park, exacerbating poverty levels. Competition for land, water, and food is rife, leading to destructive encounters between humans and animals. The respondents indicated that certain households are heavily struck by poverty as a result of threats from wild animals crossing the park, which destroys their crops, making it difficult to practice agro-based agriculture. These incidents are further clarified below:

> "..... As a result of our proximity to the Park, wild animals, particularly elephants, frequently destroy our crops. As a result, the majority of farmers no longer produce certain crops, including sorghum, which are mostly consumed by elephants. Because

the poor are getting poorer, we are not satisfied" (A male participant a Ward Councilor).

The unavailability of a compensation mechanism for victims of human–wildlife conflict was also identified as another challenge "*There is no compensation framework for human-wildlife conflict victims*" (male respondent from the Ministry of Youth, Development, Indigenization and Empowerment).

Several scholars have identified numerous challenges to PPT initiatives. For example, Muzirambi et al. (2020) pointed out that centralized governance systems adopted in community-based resource management practices cause the exclusion of local communities. This, coupled with the corruption and elite capture of PPT tourism activities, resulted in minimum benefits accruing to the locals (Shereni & Saarinen, 2021). In cases where locals perceive that they are receiving less benefits from wildlife resources, they tend to develop a negative attitude toward tourism (Tichaawa & Mhlanga, 2015). Negative perceptions are worsened by high incidents of human-wildlife conflicts and a lack of compensation for the destruction caused by wildlife (Gandiwa et al., 2013). This leads to an increase in poaching levels by community members and the formation of poaching syndicates facilitated by locals (Segobye et al., 2022). That said, it is clear that in many contexts, the poor do not benefit as much as they should from tourism, making it imperative for practitioners and policymakers to formulate effective ways to integrate the local communities fully into PPT activities (Burke, 2018).

4.4 Implications of the Study

This study has several implications for policymakers, development agencies, and other stakeholders involved in PPT. It is apparent from the results of this study that communities expect tourism to provide direct and indirect benefits that help alleviate poverty. Policymakers and local leaders must come up with viable community-based tourism enterprises managed by the locals. In addition, income-generating projects such as beekeeping and livestock farming have proven effective in reducing poverty levels. However, in the past, such projects failed because of a lack of community involvement in the management and distribution of benefits. If they are to succeed, there is a need to ensure that they are run transparently and that benefits trickle down to poor and underprivileged members of society. In addition, funding has been identified as a critical challenge that affects the viability of community-based tourism enterprises and income-generating projects. The community can partner with private investors in running community-based tourism enterprises to ensure adequate financial and human resource investments. In addition, local tourism companies, non-governmental organizations, and developmental agencies can be involved in managing and funding income-generating projects to ensure that they are viable and effective in reducing poverty levels within communities. Finally, community members must be compensated for the loss of crops, livestock, and property caused by wild animals as a result of the increase in human-wildlife conflicts. Tourism companies benefiting from nature-based tourism near protected areas should be seen to play a significant role in compensating community members and spearheading initiatives to reduce human-wildlife conflicts.

4.5 Conclusion

This study sought to ascertain the role of (PPT) in community development and its potential for poverty alleviation in communities living adjacent to the Hwange National Park. The findings of this study demonstrate the significant positive effects of this approach on the socioeconomic development of the surrounding local communities. The impact of PPT in these communities is evident through the generation of employment opportunities and the equitable distribution of tourism revenues. Residents have benefited from increased employment, leading to reduced unemployment rates and improved household incomes.

Additionally, funding for income-generating projects such as beekeeping, goat and cattle breeding, arts and crafts markets, chicken projects, and vegetable gardens has contributed to economic diversification and the retention of tourism-related revenues within the community. In terms of environmental impacts, PPT has fostered a sense of conservation and natural resource management among local communities adjacent to Hwange National Park. Through their active participation in conservation activities and revenue-sharing mechanisms, communities have become stewards of parks' natural resources, contributing to their sustainable management.

While the impact of PPT in communities adjacent to Hwange National Park has been largely positive, challenges remain. Lack of funding, inadequate benefits to the poor, and human–wildlife conflicts were the main challenges identified. Ensuring equitable benefit distribution among community members, addressing challenges brought about by human–wildlife conflicts, and balancing tourism growth with local needs require ongoing attention and collaboration among stakeholders. Continued efforts to overcome these challenges are crucial for maximizing the long-term benefits of PPT and ensuring its sustainability and effectiveness as a development strategy. Overall, the PPT has demonstrated its potential to alleviate poverty, foster sustainable development, and empower local communities adjacent to the Hwange National Park. By prioritizing community participation, equitable benefit-sharing, and sustainable practices, PPT has proven to be a valuable tool for achieving inclusive and sustainable development in communities adjacent to protected areas.

References

Ashley, C., & Goodwin, H. (2007). *'Pro-poor tourism': What's gone right and what's gone wrong?* Overseas Development Institute. www.propoortourism.org.uk/15_Gambia.pdf

Average, C. (2008). Situating Zimbabwe's community-based natural resource management practice in the sustainable development framework. *Journal of Sustainable Development in Africa, 10*(3), 172–187.

Bakker, M., & Messerli, H. R. (2017). Inclusive growth versus pro-poor growth: Implications for tourism development. *Tourism and Hospitality Research, 17*(4), 384–391. https://doi.org/10.1177/1467358416638919

Balint, P. J., & Mashinya, J. (2006). The decline of a model community-based conservation project: Governance, capacity, and devolution in Mahenye, Zimbabwe. *Geoforum, 37*(5), 805–815. https://doi.org/10.1016/j.geoforum.2005.01.011

Balint, P. J., & Mashinya, J. (2008). CAMPFIRE through the lens of the "commons" literature: Nyaminyami rural district in post-2000 Zimbabwe. *Journal of Southern African Studies, 34*(1), 127–143. https://doi.org/10.1080/03057070701832924

Braman S. 2001. Practical strategies for pro-poor tourism TROPIC Ecological Adventures - Ecuador. PPTWorking Paper Number 6.

Braun, V., & Clarke, V. (2006). Using thematic analysis in psychology. *Qualitative Research in Psychology, 3*(2), 77–101.

Burke, S. (2018). *Demystifying poverty in tourism: Looking into pro-poor tourism in India.* University of Puget Sound. https://soundideas.pugetsound.edu/ipe_theses/22/%0A

Butler, R., Curran, R., & O'Gorman, K. D. (2013). Pro-poor tourism in a first world urban setting: Case study of Glasgow Govan. *International Journal of Tourism Research, 15*, 101–113.

Cameron, T. (2018). *National parks as economic engines: An overview of economic research methods for a developing country: Case study: Hwange National Park, Zimbabwe*. Stellenbosch University.

Chambwe, M., & Saayman, A. (2023). The relationship between the agenda of conservation authorities and community development in rural areas of South Africa. *Studia Periegetica, 3*(43), 49–64.

Chambwe, M., Saayman, A., & Viviers, P. A. (2023). Rural communities' perspectives on the influence of tourism on poverty and development. *African Journal of Hospitality, Tourism and Leisure, 12*(3), 1011–1026. https://doi.org/10.46222/ajhtl.19770720.413

Child, Brian. 2003. Origins and efficacy of modern community based natural resources management (CBNRM) practices in the Southern African Region. In: Regional workshop on local communities and conservation: Issues and challenges towards a more equitable and sustainable future. : PLAAS.

Chiutsi, S., & Saarinen, J. (2017). Local participation in transfrontier tourism: Case of sengwe community in great limpopo transfrontier conservation area, Zimbabwe. *Development Southern Africa, 34*(3), 260–275. https://doi.org/10.1080/0376835X.2016.1259987

Creswell, J. W. (2013). *Qualitative inquiry & research design: Choosing among five approaches* (3rd ed.). SAGE.

Çuhadar, Y. (2024). Bibliometric analysis of studies in the context of pro-poor tourism: The case of web of science. *Journal of Tourism and Gastronomy*

Studies, 12(1), 204–226. https://doi.org/10.21325/jotags.2024.1376

Dang, P., Ren, L., & Li, J. (2024). Does rural tourism reduce relative poverty? Evidence from household surveys in western China. *Tourism Economics, 30*(2), 498–521. https://doi.org/10.1177/13548166231167648

Dube, N. (2019). Voices from the village on trophy hunting in Hwange district, Zimbabwe. *Ecological Economics, 159*, 335–343. https://doi.org/10.1016/j.ecolecon.2019.02.006

Gandiwa, E., Heitkönig, I. M. A. A., Lokhorst, A. M., Prins, H. H. T. T., & Leeuwis, C. (2013). CAMPFIRE and human-wildlife conflicts in local communities bordering northern Gonarezhou National Park, Zimbabwe. *Ecology and Society, 18*(4), 15. https://doi.org/10.5751/ES-05817-180407

Guerbois, C., Dufour, A. B., Mtare, G., & Fritz, H. (2013). Insights for integrated conservation from attitudes of people toward protected areas near Hwange National Park, Zimbabwe. *Conservation Biology, 27*(4), 844–855. https://doi.org/10.1111/cobi.12108

Jamal, T., & Dredge, D. (2014). Tourism and community development issues. In R. Sharpley & D. Telfer (Eds.), *Tourism and development* (pp. 176–204). Channel View Publications. https://doi.org/10.21832/9781845414740-008

Jones, B. T. (2004). *CBNRM, poverty reduction and sustainable livelihoods: Developing criteria for evaluating the contribution of CBNRM to poverty reduction and alleviation in Southern Africa*. PLAAS.

Jones, B. T. B., & Murphree, M. W. (2004). Community-based natural resource management as a conservation mechanism: Lessons and directions. In B. Child (Ed.), *Parks in transition biodiversity, rural development and the bottom line*. Earthscan. https://doi.org/10.4324/9781849772129

Lenao, M., Mbaiwa, J. E., & Saarinen, J. (2014). Community expectations from rural tourism development at Lekhubu Island, Botswana. Tourism Review International, 17(4), 223–236. https://doi.org/10.3727/154427214X13910101597085

Manwa, H., & Manwa, F. (2014). Poverty alleviation through pro-poor tourism: The role of Botswana Forest Reserves. *Sustainability, 6*, 5697–5713. https://doi.org/10.3390/su6095697

Marunda, E. (2014). *Primary school level and rural communities in Zimbabwe Faculty of Commerce [Univerity of Zimbabwe]*. http://ir.uz.ac.zw/bitstream/handle/10646/2687/Marunda_enhancement_of_sustainable_tourism_education_and_training.pdf?sequence=1&isAllowed=y

Matlovičová, K., Kolesárová, J., Demková, M., Kostilníková, K., Mocák, P., Pachura, P., & Payne, M. (2022). Stimulating poverty alleviation by developing tourism in marginalised roma communities: A case study of the Central Spi Region (Slovakia). *Land, 11*(10), 1–23. https://doi.org/10.3390/land11101689

Matura, P., & Mapira, J. (2018). Tourism destinations, facilities, challenges and opportunities in Zimbabwe. *European Journal of Social Sciences Studies, 2*(11), 125–138. https://doi.org/10.5281/zenodo.1217194

Medina-Muñoz, D. R., Medina-Muñoz, R. D., & Gutiérrez-Pérez, F. J. (2016). The impacts of tourism on poverty alleviation: An integrated research framework. *Journal of Sustainable Tourism, 24*(2), 270–298. https://doi.org/10.1080/09669582.2015.1049611

Mtapuri, O., & Giampiccoli, A. (2019). Tourism, community-based tourism and ecotourism: A definitional problematic. *South African Geographical Journal, 101*(1), 22–35. https://doi.org/10.1080/03736245.2018.1522598

Mutanga, C. N., Kolawole, O. D., Gondo, R., & Mbaiwa, J. E. (2024). A review and SWOC analysis of natural heritage tourism in sub-Saharan Africa. Journal of Heritage Tourism, 19(1), 49–67. https://doi.org/10.1080/1743873X.2023.2255689

Musavengane, R. (2018). Toward pro-poor local economic development in Zimbabwe: The role of pro-poor tourism. *African Journal of Hospitality, Tourism and Leisure, 7*(1), 1–14.

Musavengane, R., Siakwah, P., & Leonard, L. (2019). "Does the poor matter" in pro-poor driven sub-Saharan African cities? Towards progressive and inclusive pro-poor tourism. The International Journal of Tourism Cities, 5(3), 392–411. https://doi.org/10.1108/IJTC-05-2019-0057

Muzirambi, J. M., Musavengane, R., & Mearns, K. (2020). Revisiting devolution in community-based natural resources management in Zimbabwe: Towards inclusive governance approaches. In M. T. Stones, M. Lenao, & N. Moswete (Eds.), *Natural resources, tourism and community livelihoods in Southern Africa* (pp. 143–158). Routledge.

Njoya, E. T., & Seetaram, N. (2018). Tourism contribution to poverty alleviation in Kenya: A dynamic computable general equilibrium analysis. *Journal of Travel Research, 57*(4), 513–524. https://doi.org/10.1177/0047287517700317

Pang, Q., Hao, F., Xiao, H., & Bao, J. (2024). Community empowerment: Pro-poor tourism income distribution. Annals of Tourism Research, 106, 103764. https://doi.org/10.1016/j.annals.2024.103764

Rogerson, C. M. (2006). Pro-poor local economic development in South Africa: The role of pro-poor tourism. *Local Environment, 11*(1), 37–60. https://doi.org/10.1080/13549830500396149

Saarinen, J., & Rogerson, C. M. (2014). Tourism and the millennium development goals: Perspectives beyond 2015. *Tourism Geographies, 16*(1), 23–30. https://doi.org/10.1080/14616688.2013.851269

Saville, N. M. (2001). Practical strategies for pro-poor tourism. Case study of pro-poor tourism and SNV in Humla District, West Nepal, Pro-Poor Tourism Report, ODI, IIED, CRT

Scheyvens, R., & Momsen, J. (2008). Tourism in Small Island States: From Vulnerability to Strengths. Journal of Sustainable Tourism, 16(5), 491–510. https://doi.org/10.1080/09669580802159586

Scheyvens, R., & Hughes, E. (2019). Can tourism help to "end poverty in all its forms everywhere"? The challenge of tourism addressing SDG1. Journal of Sustainable Tourism, 27(7), 1061–1079. https://doi.org/10.1080/09669582.2018.1551404

Segobye, A. K., Mpolokang, M., Shereni, N. C., Mago, S., & Seleka, M. (2022). Community-based tourism as a pathway towards sustainable livelihoods and well-beibg in Southern Africa. In J. Saarinen, B. Lubbe, & N. N. Moswete (Eds.), *Southern African perspectives on sustainable tourism management* (pp. 125–138). Springer.

Spenceley, A. & Seif, J, (2003). Strategies, impacts and costs of pro-poor tourism approaches in South Africa, Propoor tourism working paper No. 11, Centre for Responsible Tourism, International Institute for Environment and Development, and the Overseas Development Institute.

Sharpley, R., & Naidoo, P. (2010). Tourism and poverty reduction: The case of Mauritius. Tourism and Hospitality Planning & Development, 7(2), 145–162.

Shereni, N. C. (2022). *Tourism and sustainable development goals in Zimbabwe: Contribution by the hospitality sector*. University of Johannesburg.

Shereni, N. C., & Chambwe, M. (2024). Nature-based tourism in Zimbabwe: Sustainability issues. In L. S. Stone & T. M. Stone (Eds.), *Wildlife tourism dynamics in Southern Africa: Contemporary issues, challenges and prospects for sustainable development*. Springer.

Shereni, N. C., & Saarinen, J. (2021). Community perceptions on the benefits and challenges of community-based natural resources management in Zimbabwe. *Development Southern Africa, 38*(6), 879–895. https://doi.org/10.1080/0376835X.2020.1796599

Siakwah, P., Musavengane, R., & Leonard, L. (2019). Tourism governance and attainment of the sustainable development goals in Africa. *Tourism Planning & Development, 17*, 355–383. https://doi.org/10.1080/21568316.2019.1600160

Tervo-Kankare, K., Saarinen, J., Kimaro, M. E. & Moswete, N. N. (2018). Nature-based tourism operators' responses to changing environment and climate in Uis, Namibia. African Geographical Review, 37(3), pp. 273–282.

Tichaawa, T. M., & Mhlanga, O. (2015). Community perceptions of a community-based tourism project: A case study of the CAMPFIRE programme in Zimbabwe. *African Journal for Physical, Health Education, Recreation and Dance, 21*(1), 55–67.

Truong, V. D., Hall, C. M., & Garry, T. (2014). Tourism and poverty alleviation: Perceptions and experiences of poor people in Sapa, Vietnam. *Journal of Sustainable Tourism, 22*(7), 1071–1089. https://doi.org/10.1080/09669582.2013.871019

United Nations. (2015). *Transforming our world: The 2030 agenda for sustainable development*. https://sustainabledevelopment.un.org/content/documents/21252030%20Agenda%20for%20Sustainable%20Development%20web.pdf

UNWTO. (2015). *Tourism and the sustainable development goals*. United Nations. https://doi.org/10.18111/9789284417254

Woyo, E., & Musavengane, R. (2023). Has community-based tourism been a useful tool for poverty Eradication in Southern. *African Journal of Hospitality, Tourism and Leisure, 12*(2), 521–534.

Yu, L., Wang, G., & Marcouiller, D. W. (2019). A scientometric review of pro-poor tourism research: Visualization and analysis. *Tourism Management Perspectives, 30*, 75–88. https://doi.org/10.1016/j.tmp.2019.02.005

5 Inclusion of Local Community Farmers Into the Supply Chains of Luxury Wildlife Tourism Destinations

Hugo van den Berg and Kevin Mearns

Abstract

Wildlife tourism destinations inherently demand fresh produce for memorable culinary experiences for guests and staff. Procuring local fresh produce from nearby communities can stimulate local economic development. This chapter examines a recent study on the challenges a luxury tourism destination faces when integrating local communities into its fresh produce supply chain. The study aims to assess the inclusion of local community farmers in last-mile logistics for &Beyond and its lodges: Klein's Camp in Serengeti, Mnemba Island in Zanzibar, and Forest Lodge in Phinda Private Game Reserve. A multiple-method case study research design was employed, primarily exploratory and descriptive. Data collection involved semi-structured interviews, observations, and questionnaires completed by &Beyond employees and other luxury tourism destinations. Content analysis was used for data interpretation. Findings indicate that most community farmers are family-run, and a positive association with a luxury wildlife tourism destination can inspire more families to become community farmers. Supporting community farmers generates income for them and bolsters the local economy. The study recommends that luxury wildlife tourism destinations identify potential fresh produce farmers within local communities. Once the fresh produce farmers are identified, the luxury wildlife tourism destination can inform the fresh produce farmers about the quantities of fresh produce they require.

Keywords

Luxury wildlife tourism destination · Local community farmers · Fresh produce procurement · Sustainability

5.1 Introduction

The tourism industry is one of the primary benefits of global economic growth across the world (Boesen et al., 2017; Sanches-Pereira et al., 2017). When tourists visit a wildlife tourism destination, they create a demand for local craft and fresh produce. Through local procurement, this demand creates potential opportunities for local communities by stimulating local economic development close to where communities live.

H. van den Berg (✉)
Department of Transport and Supply Chain Management, University of Johannesburg, Johannesburg, South Africa
e-mail: hugov@uj.ac.za

K. Mearns
Department of Environmental Sciences, University of South Africa, Pretoria, South Africa
e-mail: mearnkf@unisa.ac.za

K. Dube et al. (eds.), *Tourism in Rural Protected Areas*, Sustainable Development Goals Series,
https://doi.org/10.1007/978-3-031-87333-1_5

Tourism, therefore, can empower the local communities in which they live.

Tourism allows communities to develop where they live, thereby not requiring community members to seek better livelihoods in urban, mining, or industrial areas that are far removed from the community. When tourists visit a tourist destination in Africa, there will always be a demand for local craft, fresh produce, and firewood from tourists visiting the lodges, which often may be near local communities. The fundamental question is: *How can the gap between supply and demand be overcome*? Suppose a local community can realize the potential of prosperity lying dormant within the tourism supply chain (SC); in the long run, the financial and social benefits will be well worth it for the local community (Rylance et al., 2017). Community involvement in logistics and supply chain-related processes of tourism destinations is perceived as not very important by supply organizations. However, it is estimated that 75% of a company's carbon footprint is produced by its supply chain (Van den Berg et al., 2013).

Tourism companies should seriously investigate and consider their suppliers and supply chains to reduce their environmental footprints. According to Lukhele and Mearns (2013), changes in the tourism industry are unavoidable and ever-changing and should be embraced by tourism organizations. When a wildlife tourism destination can establish a working relationship with community farmers, it contributes to sustainability and green procurement practices (GPP). GPP is the process of procuring fresh produce from local community farmers, considering the social, economic, and environmental impacts of the local community. When local communities are included in the last-mile distribution of fruit and vegetable supply chains (SCs) for wildlife tourism establishments, they could have significant financial and social benefits for the local communities.

Local communities rarely benefit from tourism; however, incorporating local communities that produce fresh produce into the luxury wildlife tourism destination supply chain could present an opportunity for local communities to benefit (Rylance et al., 2017). However, little information is available to support such collaboration. Therefore, it is difficult to determine whether local communities producing fresh produce can be incorporated into the supply chain of luxury wildlife tourism destinations, as there is a lack of information available to support such collaboration (Arampantzi & Minis, 2017). However, Dodds et al. (2016) state that there is sufficient research focused on local communities and their needs and how tourism is valued as a tool for community conservation and development.

According to Dodds et al. (2016), this effort is known as community-based tourism (CBT) when local communities engage in tourism. Community-based tourism is, therefore, a mechanism that can help local communities develop opportunities to improve their living conditions over the long term. The goal of CBT is to achieve sustainable community development. Thus, it can be considered an alternative form of tourism (Rylance et al., 2017). From tourism research, it was assumed that the mindset of a local community must be positive toward supplying locally produced fresh produce to luxury wildlife tourism destinations (McCamley & Gilmore, 2017). Likewise, the objectives of providing and receiving locally produced fresh produce from both the local community and the luxury wildlife tourism destination point of view must focus on creating a successful partnership while ensuring customer satisfaction (Boesen et al., 2017). Collaboration can sometimes be problematic because the local community's viewpoints, interests, and goals may differ from those of luxury wildlife tourism destinations (McCamley & Gilmore, 2017).

Both luxury tourism destinations and local communities can benefit from fresh produce being provided to luxury tourism destinations. Tourists are able to taste and consume local fresh produce at mealtimes. Tourists can visit local community farmers to experience the environment where these community farmers function (Ndivo & Cantoni, 2015). Taste experience is related to the personal, social, and cultural background of tourists visiting a luxury tourism destination (Anderson et al., 2017). Anderson et al.

(2017) stated that numerous scholars have thoroughly researched food production and hospitality. However, only recently has "food tourism" gained prominence in the study of tourism and food production. Through tourism, food tourism must be careful not to see local communities as *tourism for the poor*, but as humanitarian initiatives focused on the local communities, making tourists aware of their obligations to support local communities.

The purpose of this chapter is to provide an overview of the challenges a luxury tourism destination is exposed to when it considers the inclusion of a local community in its fresh produce SC. This chapter highlights a framework in which luxury wildlife tourism destinations can consider maximizing local community contributions to their last-mile SC processes. The information in this chapter is presented on a characteristic-by-characteristic basis. The described characteristics were selected according to the current and future functionality of the SC processes of fresh produce for luxury wildlife tourism destinations.

5.2 Tourism Supply Chains

COVID-19 has tremendously affected the global travel industry, with a worldwide decline of 90% (Bian et al., 2021; Dube & Nhamo, 2024). Some tourism destinations were forced to reduce their workforce by 70%, and a substantial number of tourism destinations were unlikely to generate profits because of the COVID-19 pandemic in 2020 (Gursoy & Chi, 2020; Rizou et al., 2020). However, this situation has changed since 2023, and the tourism industry is returning to its pre-COVID-19 levels. Tourists visiting tourism destinations will increase their demand for local crafts and fresh produce. The fundamental question is: How can the gap between the supply and demand of local craft and fresh produce be overcome? Can local communities realize the potential prosperity that lies dormant within the tourism SC? In this case, the financial and social benefits will be worth it for the local community (Touboulic & Walker, 2015; Spenceley et al., 2019;).

Much research has been conducted on Supply Chain Management (SCM) and SC operations, and many similar problems and difficulties have been identified in these SCs. However, the development of tourism SCs is a relatively new concept and noticeable problems and challenges within tourism SCs are surfacing (Michailidou et al., 2016; Rao, 2017). In an SC milieu, there is no standard model that will work for similar types of companies, even if these companies operate in the same business environment (Botes et al., 2017; Elgazzar et al., 2019). Very little research has been conducted on the inclusion of local communities in last-mile logistics systems has been done. According to Shahbaz et al. (2018), organizations are working more closely to incorporate upstream and downstream SC partners. Innovation in SCs is sometimes difficult for organizations to explore and implement for the following two reasons. Firstly, internal organizational structures, systems, and commitment are barriers that need to be addressed or overcome. Secondly, external factors such as knowledge sharing, lack of resources, financial difficulties, unstable demand, and commitment hinder SC innovation. Research has shown that the concept of SC innovation is an area in the SCM process that has not received a great deal of attention.

Supply chain managers in organizations are continuously searching for innovations to bring into their SC to satisfy their customers' needs or achieve a competitive advantage, which will be long-lasting and problematic to duplicate. The concept of innovative supply chain practice is to discover and implement an idea, notion, or system between SC partners and/or organizations that were of a problematic nature or did not transpire before (Dubey et al., 2017). Regarding SC lead-time reduction, the question of whether the inclusion of local suppliers in the SC will shorten the SC lead-time is questionable.

In recent years, regional and international organizations across all SCs domains have started to involve local communities in some of their SC operations to determine whether cost savings or product substitution is possible (Pooe & Mahlangu, 2017; Skippari et al., 2017). The costs (technical and financial support) involved in including local communities in SCs do not necessarily justify the cost savings obtained (Bienhaus & Haddud, 2018). Why, then, do local and international organizations support local communities? Rylance and Spenceley (2014) argue that the reason why some tourism organizations involve the local community to become part of their SC is motivated more by their business ethic, and seems to be the "right thing to do" rather than a desire to see a return on investment (Estol et al., 2018; Mani & Gunasekaran, 2018). Supply chain development in an organization focuses on the SC's ability or competencies to deliver the desired SC outcomes for the organization.

The notion of SC sustainability looks at the practices and actions needed at the micro level that contribute to or hinder last-mile logistics distribution system sustainability (Bi et al., 2017; Da Silva et al., 2019; Koberg & Longoni, 2019). The focus has moved beyond the economic dimensions of development to encompass the social and cultural aspects of how communities can benefit from last-mile logistic distribution system sustainability. This research aims to firstly establish the fresh produce demands of the &Beyond lodges. Secondly, we determined the fresh produce supply capabilities of the local communities. Thirdly, to assess the latent, new, and present possibilities of the possible inclusion of local community farmers into the last-mile logistic distribution systems of selected &Beyond lodges. Luxury tourism destination companies, such as &Beyond, are under significant pressure because clients visiting luxury tourism destinations in Sub-Saharan Africa expect world-class accommodation and services.

This study aims to determine the possibility of including fresh produce from local community farmers in last-mile logistics distribution systems within a tourism case study company, namely &Beyond. The purpose of assessing the possibility of including fresh produce from local farmers is to contribute to community development and the creation of a sustainable fresh produce SC. &Beyond can play an essential role in developing and assembling knowledge regarding a sustainable fresh produce supply within the company, which can assist in achieving a competitive advantage in their fresh produce SC (Biao et al., 2014; Ndivo & Cantoni, 2015; Raj et al., 2018). The incorporation of local communities into the SC could facilitate a luxury wildlife tourism destination to achieve the sustainable development goals (SDG) UNSDG1—no poverty. The SC of products, including fresh produce, is complex for the &Beyond lodges. The inherent uncertainty of the tourism environment and SCs contributes to the complexity of restructuring the fresh produce SC (Abbasi, 2017).

In the tourism industry, luxury wildlife tourism destinations have different SCs that supply a diverse range of products to various camps and lodges within these SCs, which are the lifelines of the business processes of different camps and lodges. Gattorna (2015) suggested that companies must align their SCs strategies to focus on customer needs. This means that tourism companies such as and &Beyond should try to incorporate the views and needs of their customers into their product offerings. The inclusion aligns closely with the needs of tourists for local communities to receive a greater portion of benefits. &Beyond operates 29 camps and lodges in spectacular locations, from Serengeti and the Okavango Delta to the islands of the Indian Ocean. The inclusion of all lodges in the research process would have been a considerable undertaking. Therefore, for the research process, only three &Beyond lodges, (1) Klein's Camp in Serengeti, (2) Mnemba Island in Zanzibar, and (3) Forest Lodge in Phinda Private Game Reserve, were used (Fig. 5.1).

These three &Beyond lodges were selected because they are located in different and unique environments. These three lodges provide opportunities for unique SC challenges. Hence, the data collection through questionnaires, personal interaction with community farmers, and telephonic interviews with key people for other luxury wildlife tourism destinations were used to collect a diverse range of data used in the case study research.

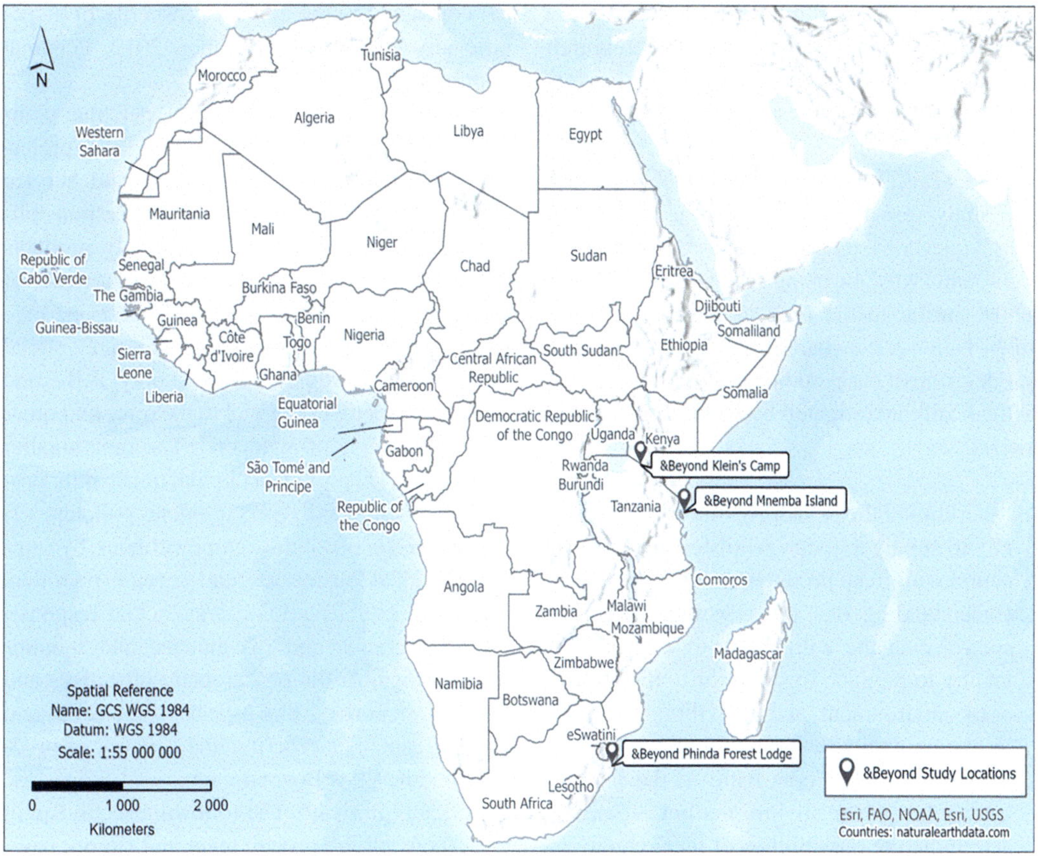

Fig. 5.1 &Beyond lodges study sites. Source: Esri, FAO, NOAA, USGS, naturalearthdata.com

5.3 Methodology

A multiple-method case study was the primary research design for this study. According to Yin (2018, p. 18), "the essence of a case study, the central tendency among all types of case study, is that it tries to illuminate a decision or set of decisions: why they were taken, how they were implemented, and with what result." The research questions in this case study led to several exploratory and descriptive case studies. In this chapter, several exploratory and illustrative case studies form analytical weights. Therefore, the research method in this chapter is primarily exploratory and descriptive in nature. This study aims to establish whether it is possible to include fresh produce from local communities in the last-mile logistics distribution systems of luxury wildlife tourism destinations. When using case studies, the researcher develops an understanding of processes, events, or programs of situations and activities in various environments.

In mixed-method research, three different designs can be used: (1) convergent parallel mixed methods, (2) explanatory sequential mixed methods, and (3) exploratory sequential mixed methods (Creswell & Creswell, 2017). These case studies can assist the researcher in collecting information using different collection procedures to develop an outcome for a research objective. Case studies are used to explore something new or to investigate a particular scenario rather than a hypothesis being examined (Mligo, 2016). In this chapter, case study research and a convergent parallel mixed method were used as the primary research strategy.

Case study research is demarcated in different ways, and its focus of case study research can be on a single case or multiple similar variable cases (Patten & Newhart, 2018). In case study research, researchers attempt to comprehensively investigate and explain single or multiple occurrences, taking into account as many considerations as possible (Tan, 2017). Yin (2018) elaborates that case study research focuses on the "how" and "why" questions. The researcher can determine "how" and "why" by using case studies that cover or are similar to the outcomes of the research study (Patten & Newhart, 2018). The exploratory and descriptive case studies in this chapter were methodically investigated based on the following criteria.

- Determining the capability of a local community to supply a luxury wildlife tourism destination with fresh produce.
- Understanding that the researcher has little control over the willingness of a local community to produce fresh produce, the weather or the environment, and the willingness of the local community to support.
- This research can contribute to the body of knowledge when all information is used to determine the possibilities of local communities becoming part of the last-mile logistics distribution systems of FMCG for luxury wildlife tourism destinations.

Secondary data analysis (SDA) in the form of actual &Beyond orders from the three selected lodges were analyzed to determine the procurement of fresh produce. A Preferential Procurement Policy (PPP) was implemented by &Beyond in July 2018. PPP encourages the active participation of &Beyond's business partners to establish a broader and meaningful business relationship. Through business relationships, &Beyond's vision is to: (1) help create economic growth within a specific area/community; (2) promote sustainable development of the business partners within a specific area/community; (3) contribute to job creation within a specific area/community; (4) establish the possibility of new business opportunities in a specific area/community, and (5) contribute to the general prosperity of a specific area/community (Brenner, 2019, Personal communication).

According to Pretorius (2019, personal communication), when implementing PPP, preference is given firstly to suppliers and service providers who have corporate social responsibility programs in place and can supply products equal to the price and quality requirements of &Beyond. Secondly, suppliers and service providers without established corporate social responsibility programs can supply &Beyond with products that are equal to the price and quality requirements of &Beyond. The functionality of PPP is to (1) provide a fair and transparent process for &Beyond's procurement policies; (2) contribute to economic empowerment by supporting local businesses and service providers who have established corporate social responsibility programs; and (3) manage and monitor progress against the procurement objectives and regularly report on this aspect by the &Beyond impact team. The ordering and delivery of products for the 29 &Beyond camps and lodges are remarkably intricate. The following sections will only focus on the procurement and supply pathways of (1) Forest Lodge in Phinda Private Game Reserve, (2) Klein's Camp in the Serengeti, and (3) Mnemba Island in Zanzibar.

In this chapter, various research methods are combined. First, to study (through observation, questionnaires, and interviews) a few communities close to the selected luxury wildlife tourism destinations and to determine whether local produce is grown in the community. Second, interviews with key informants were conducted to determine the procurement pathway of products, including fresh produce, for three selected &Beyond lodges. Different research methods were developed specifically to address the research objectives. The SDA is supported by researchers' participant observations in the communal areas near the Lodges as well as interviews with 60 community farmers to determine what kind of local produce is grown within the applicable communities.

5.4 Results and Discussion

The mission of &Beyond is to provide customers visiting luxury wildlife tourism destinations with memorable and unique culinary experiences. To achieve this, a team of culinary experts developed several 7 days of lunch, dinner, and high tea menus in &Beyond the head office in Johannesburg (Jardine, 2020, Personal communication). These menus are sent to all the &Beyond Lodges. Lodge managers and head chefs use these menus as a basis to plan different meals and high tea servings for the guests staying at the lodges. Each lodge is free to add any type of food amenity to its menus to elevate the dining experience and the quality of the serving or to accommodate a specific food requirement for a guest. All the &Beyond Lodges send orders to a particular &Beyond office. Once these orders are captured, various articles and food products are procured and distributed to &Beyond Lodges.

The lodge managers from the three &Beyond Lodges are responsible for submitting their orders to the &Beyond offices. The Phinda Private Game Reserve sends its orders to the &Beyond office in Johannesburg, Mnemba Island sends its orders to the &Beyond offices in Stonetown in Zanzibar, and Klein's Camp sends their orders to the &Beyond offices in Arusha, which is in Tanzania (Brenner, 2019, June 19). The orders received from lodges are for a diverse range of products. Some products are kept in &Beyond offices in Johannesburg, Arusha, and Zanzibar. If a product is unavailable at the &Beyond offices of Johannesburg, Arusha, and Zanzibar, a new order will be sent to the supplier/s of that particular product.

5.4.1 Fresh Produce Requirements for &Beyond Lodges

The &Beyond Lodges order various products based on the specific requirements of a particular lodge. This section determines the supply and demand of fresh produce for Klein's Camp, Mnemba Island, and the Phinda Forest Lodge. The &Beyond lodge managers of the three lodges provided researchers with many completed orders. It must be noted that, Firstly, the orders which were examined did not include every order sent by Klein's Camp, Mnemba Island, and Phinda Forest Lodge to the &Beyond offices over the past few years because some of the orders were irrelevant to this research. This is because fresh produce was not a part of some orders. Secondly, all orders obtained from the &Beyond Lodges contained detailed information that was considered sensitive. Lastly, some purchasing orders received were not official internal &Beyond orders but were copies of invoices issued by various retail companies to some of the &Beyond lodges. These orders included various products, including fresh produce, which retail companies delivered directly to &Beyond Lodges. An estimate of the demand for fresh produce needed to be determined for Klein's Camp, Mnemba Island, and Phinda Forest Lodge (Table 5.1). This was achieved after examining the official internal &Beyond orders, as well as the invoices from the retail companies of Klein's Camp, Mnemba Island, and Phinda Forest Lodge from 2019 to 2021.

Information from the orders provided an overview of the quantities of fresh produce used at Klein's Camp, Mnemba Island, and Phinda Forest Lodge. It is important to identify the demand for fresh produce for each lodge, as it provides an overall view of the orders placed on the different fresh produce. The focus of the research was to identify the different types, not the quantities, of fresh produce used by the three lodges. Therefore, it was difficult for guests and staff to determine the volumes of different types of fresh produce. This would, however, be important for future research to establish the volume of fresh produce required for guests and staff. Luxury wildlife tourism destinations, such as &Beyond, can use this information to create working relationships with fresh produce suppliers to safeguard themselves against potential future supply chain disruptions.

Table 5.1 Fresh produce demand of Klein's Camp, Mnemba Island, and Phinda Forest Lodge

Fresh produce	Klein's Camp	Mnemba island	Phinda Forest Lodge	Fresh produce	Klein's Camp	Mnemba island	Phinda Forest Lodge	Fresh produce	Klein's Camp	Mnemba island	Phinda Forest Lodge
Apples - Green	✓	✓	✓	Green Beans	✓	✓	✓	Papaya	✓	✓	✓
Arrowroot (Magimbe)	✓	✓	✓	Granadilla	✓	✓	✓	Paw-Paw	✓	✓	✓
Asparagus	✓	✓	✓	Grapefruit	✓	✓	✓	Pears - Pacham	✓	✓	✓
Aubergine	✓	✓	✓	Guava	✓	✓	✓	Pears - Sugar Snap	✓	✓	✓
Avocadoes	✓	✓	✓	Kiwi fruit	✓	✓	✓	Pepper, Green	✓	✓	✓
Bananas	✓	✓	✓	Lemon	✓	✓	✓	Pepper, Red	✓	✓	✓
Beetroot	✓	✓	✓	Lettuce	✓	✓	✓	Pineapples	✓	✓	✓
Broccoli	✓	✓	✓	Lime	✓	✓	✓	Potatoes, Iris	✓	✓	✓
Butternut	✓	✓	✓	Maize	✗	✗	✓	Pumpkin	✓	✓	✓
Brown Lentils	✓	✓	✓	Mango	✓	✓	✓	Raspberries	✓	✓	✓
Cabbage	✓	✓	✓	Mangoes	✓	✓	✓	Strawberries	✓	✓	✓
Carrot	✓	✓	✓	Melon, Sweet	✓	✓	✓	Spinach	✓	✓	✓
Cauliflower	✓	✓	✓	Naartjies	✓	✓	✓	Spinach, Wild	✓	✓	✓
Chillies	✓	✓	✓	Oranges	✓	✓	✓	Sweet Potatoes	✓	✓	✓
Corn - Baby	✓	✓	✓	Onion, Red	✓	✓	✓	Tomatoes	✓	✓	✓
Cucumber	✓	✓	✓	Onion, Spring	✓	✓	✓	Tomatoes, Cherry	✓	✓	✓
Garlic - Fresh	✓	✓	✓	Onion, White	✓	✓	✓	Watermelon	✓	✓	✓
Ginger - Fresh	✓	✓	✓	Paw-Paw	✓	✓	✓	Zucchini	✓	✓	✓
Gem squash	✓	✓	✓	Passion Fruit	✓	✓	✓				
	19	19	19		18	18	19		19	19	19

The demand was 56 types of fresh produce between the three &Beyond Lodges
Klein's Camp demand was for 55 of the 56 types of fresh produce
Mnemba Island's demand was for 55 of the 56 types of fresh produce
Phinda Forest's demand was for 56 of the 56 types of fresh produce

5.4.2 Fresh Produce Supply Capabilities of Local Communities Around &Beyond Lodges

The previous section provided an essential context for the use of fresh produce for Klein's Camp, Mnemba Island, and Phinda Forest Lodge. This requires an understanding of the demands for different types of fresh produce. The purpose of this section is to identify the types of fresh produce local communities around Klein's Camp, Mnemba Island, and Phinda Forest Lodge are currently producing. The relevance and possible importance of this information regarding the supply of fresh produce to lodges by local farmers were investigated. A total of 60 local community farmers (30 around Mnemba Island and 30 close to Phinda Forest Lodge) were visited to determine the different types of fresh produce these community farmers were cultivating. As described in the section above, Klein's Camp is the only lodge with one local community, a Maasai village located approximately 10 km away. At Klein's, there were no community farmers in the village. In 2015, staff from Klein's Camp assisted the people from Maasai village in developing a vegetable garden, known as a 'shamba,' according to the Maasai people; after nine months, the shamba produced 11 different kinds of fresh produce for Klein's Camp and the people within Maasai village (Strautmann, 2020, Personal communication).

Because of COVID-19-related travel restrictions, the authors could not visit Klein's camp, Mnemba Island, or Phinda Forest Lodge after preliminary site visits. Fortunately, two local &Beyond employees at Mnemba Island and Phinda Forest Lodge assisted the authors in data collection. The lodge manager of Klein's Camp, Mr. Strautmann, collected data from the 'shamba' near Klein's Camp. Unfortunately, because of the COVID-19 pandemic, tourists could not visit Klein's Camp; therefore, the demand for fresh

Table 5.2 Fresh produce produced by local farmers

Fresh produce	Shamba near Klein's Camp	Farmers around Mnemba Island	Farmers around Phinda	Fresh produce	Shamba near Klein's Camp	Farmers around Mnemba Island	Farmers around Phinda	Fresh produce	Shamba near Klein's Camp	Farmers around Mnemba Island	Farmers around Phinda
Apples - Green	✗	✗	✗	Green Beans	✗	✗	✓	Papaya	✗	✗	✓
Arrowroot (Magimbe)	✗	✗	✗	Granadilla	✗	✗	✓	Paw-Paw	✗	✗	✗
Asparagus	✗	✗	✗	Grapefruit	✗	✓	✗	Pears - Pacham	✗	✗	✗
Aubergine	✓	✓	✗	Guava	✗	✗	✗	Pears - Sugar Snap	✗	✗	✗
Avocadoes	✓	✗	✓	Kiwi fruit	✗	✓	✗	Pepper, Green	✗	✓	✓
Bananas	✗	✗	✓	Lemon	✓	✗	✗	Pepper, Red	✗	✗	✓
Beetroot	✓	✗	✓	Lettuce	✗	✗	✓	Pineapples	✗	✓	✗
Broccoli	✓	✗	✓	Lime	✗	✓	✗	Potatoes, Iris	✗	✗	✓
Butternut	✗	✗	✓	Maize	✓	✓	✓	Pumpkin	✓	✓	✓
Brown Lentils	✗	✗	✓	Mango	✗	✓	✓	Raspberries	✗	✗	✗
Cabbage	✓	✗	✓	Mangoes	✗	✓	✓	Strawberries	✗	✗	✗
Carrot	✗	✗	✓	Melon, Sweet	✗	✗	✗	Spinach	✓	✓	✓
Cauliflower	✗	✗	✓	Naartjies	✗	✗	✓	Spinach, Wild	✗	✗	✓
Chillies	✗	✓	✓	Oranges	✗	✗	✓	Sweet Potatoes	✓	✓	✓
Corn - Baby	✗	✗	✗	Onion, Red	✗	✗	✓	Tomatoes	✓	✓	✓
Cucumber	✗	✗	✗	Onion, Spring	✗	✗	✓	Tomatoes, Cherry	✗	✗	✗
Garlic - Fresh	✗	✗	✗	Onion, White	✗	✗	✓	Watermelon	✗	✓	✓
Ginger - Fresh	✗	✗	✗	Paw-Paw	✗	✓	✗	Zucchini	✗	✗	✗
Gem squash	✗	✗	✓	Passion Fruit	✗	✗	✗				
	5	3	11		2	7	12		4	7	10

Shamba near Klein's Camp cultivated 11 of the 56 types of fresh produce
Farmers around Mnemba Island cultivated 17 of the 56 types of fresh produce
Farmers around Phinda cultivated 33 of the 56 types of fresh produce

produce declined, which resulted in the shamba becoming obsolete and forced to shut down in 2020 (Strautmann, 2020, Personal communication). The 'shamba' has, however, been reopened in 2022 and is again producing fresh produce for Klein's Camp. After the data were collected and examined, the information was used to provide an overview (Table 5.2) of the fresh produce produced by local farmers around Mnemba Island and Phinda Forest Lodge.

The information has provided an overview of the fresh produce the local community farmers, and the "shamba," were currently producing in 2020. It was important to identify the types of fresh produce that communities could supply in order to provide an overall view of what was available. However, it is clear that community farmers around Mnemba, on average, produce only 17 of the 56 different types of fresh produce required by Mnemba Island. On average, community farmers around Phinda Forest Lodge produce 33 of the 56 different types of fresh produce required by Phinda Forest Lodge. In Tanzania, the "shamba" can only produce 11 of the 56 different types of fresh produce required by Klein's Camp. Therefore, it is evident that community farmers produce only a small number of fresh produce types that can contribute to the demand requirements of Klein's Camp, Mnemba Island, and Phinda Forest Lodge. This section laid the foundation for the ordering processes of Klein's Camp, Mnemba Island, and Phinda Forest Lodge. The demand for different types of fresh produce is also highlighted. Furthermore, it was determined that different types of fresh produce could be produced and supplied by community farmers close to the Phinda Forest Lodge, Mnemba Island, and the shamba near Klein's Camp. It has always been an important endeavor to provide guests visiting any of the &Beyond lodges with a memorable culinary experience (Jardine, 2021, Personal communication). Good quality fresh produce must be used to satisfy guest requirements.

5.5 Conclusion and Recommendations

The demand for different types of fresh produce in Klein's Camp, Mnemba Island, and Phinda Forest Lodge has been highlighted. In addition, different types of fresh produce can be produced by community farmers close to Phinda Forest Lodge, Mnemba Island, and the "shamba" near Klein's Camp. This highlights the ability of community farmers to supply fresh produce to Klein's Camp, Mnemba Island, and Phinda Forest Lodge. This study aimed to determine whether it would be possible to include fresh produce from community farmers in the last-mile logistics distribution systems of luxury wildlife tourism destinations. It was established through research that the majority (76%) of luxury wildlife tourism destinations indicated that lodges would consider buying fresh produce. Thus, it is recommended that the luxury wildlife tourism destinations identify potential fresh produce farmers in the local community. Once the fresh produce farmers are identified, the luxury wildlife tourism destination can inform them, fresh produce farmers, about the quantities of fresh produce they require. The luxury wildlife tourism destination could work closely with community farmers to assist in developing a greater variety of fresh produce, which is required by the luxury wildlife tourism destination. Luxury wildlife tourism destinations in a certain area could also work together to create a social media platform that will list the community farmers in a certain area and the produce they have available. The platform can then be accessed by other lodges, the local community, and the public to locally produce sources (Van den Berg & Mearns, 2021).

This study aimed to identify the possibilities of including fresh produce from community farmers in the last-mile SC processes of luxury wildlife tourism destinations. This emphasizes the need for a holistic approach to the essential requirements to establish a strategic framework to determine the probabilities for the inclusion of community farmers in the last-mile SC processes of luxury wildlife tourism destinations. Many organizations worldwide that are directly or indirectly related to SCs are working together to identify and rationalize the full range of SC impediments these organizations face. To successfully deal with these obstacles, luxury wildlife tourism destinations can do one of the following: first, complete analysis of their current product SCs, including fresh produce. Second, we obtained information regarding alternative suppliers, focusing on local community farmers as a potential business opportunity. Last, luxury wildlife tourism destinations must implement a strategic framework for the possible inclusion of fresh produce from community farmers as part of a long-term strategy to create an alternative sustainable fresh produce supply chain.

References

Abbasi, M. (2017). Towards socially sustainable supply chains–themes and challenges. *European Business Review, 29*(3), 261–303.

Anderson, T. D., Mossberg, L., & Therkelsen, A. (2017). Food and tourism synergies: Perspectives on consumption, production and destination development. *Scandinavian Journal of Hospitality and Tourism, 17*(1), 1–8.

Arampantzi, C., & Minis, I. (2017). A new model for designing sustainable supply chain networks and its application to a global manufacturer. *Journal of Cleaner Production, 156*(6), 276–292.

Bi, R., Davison, R. M., & Smyrnios, K. X. (2017). E-Supply chain coordination and SME performance: An empirical investigation. *Electronic Journal of Information Systems Evaluation, 20*(2), 1–16.

Bian, Z., Zuo, F., Gao, J., Chen, Y., Venkata, S. S. C. P., Bernardes, S. D., Ozbay, K., Ban, X. J., & Wang, J. (2021). Time lag effects of COVID-19 policies on transportation systems: A comparative study of New York City and Seattle. *Transportation Research Part A: Policy and Practice, 145*(3), 269–283.

Biao, L., Liang, W., & Liang, L. (2014). A multi-agent-based research on tourism supply chain risk management. *Journal of Advanced Manufacturing Systems, 13*(3), 133–153.

Bienhaus, F., & Haddud, A. (2018). Procurement 4.0: Factors influencing the digitisation of procurement and supply chains. *Business Process Management Journal, 24*(4), 965–984.

Boesen, M., Sundbo, D., & Sundbo, J. (2017). Local food and tourism: An entrepreneurial network approach. *Scandinavian Journal of Hospitality and Tourism, 17*(1), 76–91.

Botes, A., Niemann, W., & Kotze, T. (2017). Buyer-supplier collaboration and supply chain resilience: A case study in the petrochemical industry. *South African Journal of Industrial Engineering, 28*(4), 183–199.

Creswell, J. W., & Creswell, J. D. (2017). *Research design: Qualitative, quantitative and mixed methods approaches* (4th ed.). SAGE Publications.

Da Silva, F. P., Brandão, F., & Sousa, B. (2019). Towards socially sustainable tourism in cities: Local community perceptions and development guidelines. *Enlightening Tourism: A Pathmaking Journal, 9*(2), 168–198.

Dodds, R., Ali, A., & Galaski, K. (2016). Mobilizing knowledge: Determining key elements for success and pitfalls in developing community-based tourism. *Current Issues in Tourism, 21*(13), 1547–1568.

Dube, K., & Nhamo, G. (2024). Tourism resilience and challenges in Limpopo, South Africa: A post-COVID-19 analysis. *Development Southern Africa, 41*, 686–703.

Dubey, R., Gunasekaran, A., Papadopoulos, T., Childe, S. J., Shibin, K. T., & Wamba, S. F. (2017). Sustainable supply chain management: Framework and further research directions. *Journal of Cleaner Production, 142*(2), 1119–1130.

Elgazzar, S., Tipi, N., & Jones, G. (2019). Key characteristics for designing a supply chain performance measurement system. *International Journal of Productivity and Performance Management, 68*(2), 296–318.

Estol, J., Camilleri, M. A., & Font, X. (2018). European Union tourism policy: An institutional theory critical discourse analysis. *Tourism Review, 73*, 421–431.

Gattorna, J. (2015). *Dynamic supply chains. How to design, build and manage people-centric value networks* (3rd ed.). Pearson Education.

Gursoy, D., & Chi, C. G. (2020). Effects of COVID-19 pandemic on hospitality industry: Review of the current situations and a research agenda. *Journal of Hospitality Marketing & Management, 29*(5), 527–529.

Koberg, E., & Longoni, A. (2019). A systematic review of sustainable supply chain management in global supply chains. *Journal of Cleaner Production, 207*(1), 1084–1098.

Lukhele, S., & Mearns, K. (2013). The operational challenges of community-based tourism ventures in Swaziland. *African Journal for Physical, Health Education, Recreation and Dance, 2*, 199–216.

Mani, V., & Gunasekaran, A. (2018). Four forces of supply chain social sustainability adoption in emerging economies. *International Journal of Production Economics, 199*(5), 150–161.

McCamley, C., & Gilmore, A. (2017). Aggravated fragmentation: A case study of SME behaviour in two emerging heritage tourism regions. *Tourism Management, 60*(1), 81–91.

Michailidou, A. V., Vlachokostas, C., Achillas, C., Maleka, D., Moussiopoulos, N., & Feleki, E. (2016). Green tourism supply chain management based on life cycle impact assessment. *European Journal of Environmental Sciences, 6*(1), 30–36.

Mligo, E. S. (2016). *Introduction to research methods and report writing. A practical guide for researchers in social sciences and humanities*. World Scientific Publishing Co.

Ndivo, R. M., & Cantoni, L. (2015). Economic empowerment of communities through tourism: A pro-poor tourism value chain approach. *Advances in Hospitality and Tourism Research, 3*(2), 116–134.

Patten, M. L., & Newhart, M. (2018). *Understanding research methods. An overview of essentials* (10th ed.). Routledge.

Pooe, R. I. D., & Mahlangu, D. (2017). Enhancing SME performance through supply chain integration, collaborative planning, and supply chain capabilities – research. *Journal of Contemporary Management, 14*(1), 238–269.

Raj, A., Biswas, I., & Srivastava, S. K. (2018). Designing supply contracts for the sustainable supply chain using game theory. *Journal of Cleaner Production, 185*(6), 275–284.

Rao, K. (2017). Purchasing behaviour of rural consumers in South India with reference to FMCG. *International Journal of Research in Commerce & Management, 8*(8), 2–18.

Rizou, M., Galanakis, I. M., Aldawoud, T. M., & Galanakis, C. M. (2020). Safety of foods, food supply chain and environment within the COVID-19 pandemic. *Trends in Food Science & Technology, 102*(8), 293–299.

Rylance, A., Snyman, S., & Spenceley, A. (2017). The contribution of tourism revenue to financing protected area management in Southern Africa. *Tourism Review International, 21*(2), 139–149.

Rylance, A., & Spenceley, A. (2014). Living outside the fence: opportunities for neighbouring communities to supply products and services to the Sabi Sand Game Reserve, South Africa. *African Journal for Physical Health Education, Recreation and Dance, 19*(3), 294–308.

Sanches-Pereira, A., Onguglo, B., Pacini, H., Gómez, M. F., Coelho, S. T., & Muwang, M. K. (2017). Fostering local sustainable development in Tanzania by enhancing linkages between tourism and small-scale agriculture. *Journal of Cleaner Production, 162*(6), 1567–1581.

Shahbaz, M. S., Rasi, R. Z. R., Ahmad, M. B., & Sohu, S. (2018). The impact of supply chain collaboration on operational performance: Empirical evidence from manufacturing of Malaysia. *International Journal of Advanced and Applied Sciences, 8*, 64–71.

Skippari, M., Laukkanen, M., & Salo, J. (2017). Cognitive barriers to collaborative innovation generation in supply chain relationships. *Industrial Marketing Management, 62*(3), 108–117.

Spenceley, A., Snyman, S., & Rylance, A. (2019). Revenue sharing from tourism in terrestrial African protected areas. *Journal of Sustainable Tourism, 27*(6), 720–734.

Tan, W. (2017). *Research methods: A practical guide for students and researchers*. World Scientific Publishing.

Touboulic, A., & Walker, H. (2015). Theories in sustainable supply chain management: A structured literature review. *International Journal of Physical Distribution and Logistics Management, 45*(1/2), 16–42.

Van den Berg, H., & Mearns, K. (2021). An exploration of guesthouse fresh produce purchasing behaviour and supply chains in Johannesburg. *Journal of Transport and Supply Chain Management, 15*, 1–8.

Van den Berg, U., Labuschagne, J. P., & Van den Berg, H. (2013). The effects of greening the supplier and innovation on environmental performance and competitive advantage. *Journal of Transport and Supply Chain Management, 7*(1), 1–7.

Yin, R. K. (2018). *Case study research and applications. Design and methods* (6th ed.). SAGE Publications.

Personal Communication

Brenner, B. Logistics Manager, & Beyond (2019). Personal communication 6 July, Johannesburg.

Jardine, C. General Manager, & Beyond (2020). Personal communication 12 February, Phinda Forest, Vlei and Homestead Lodges.

Jardine, C. General Manager, & Beyond (2021). Personal communication 19 October, Phinda Forest, Vlei and Homestead Lodges.

Pretorius, K. Procurement Manager, & Beyond (2019). Personal communication 4 August, Johannesburg.

Strautmann, K. Lodge Manager, & Beyond (2020). Personal communication 8 March, Tanzania.

Capability Development for Rural Tourism: Communities Around a Protected Area

6

Nelisile Zanele Mgabhi,
Ikechukwu O. Ezeuduji,
and Nomfundo Sinethemba Mzimela

Abstract

Rural tourism may benefit rural communities that generally depend on traditional livelihood activities such as farming, fishing, and mining. However, the preconditions, skills, and capabilities for rural tourism development need to be researched before local communities embark on such development. This chapter presents a recent study that explored the rural tourism preconditions, skills, and capabilities of local communities neighboring a protected area, namely, Hluhluwe iMfolozi Park, in KwaZulu-Natal Province, South Africa, for rural tourism development. This study adopted a qualitative research method that hinges on explorative and interpretive paradigms. Data were collected through semi-structured interviews and focus group discussions with the participants. Content analysis was used to analyze the data.

Results: The study findings highlight the internal and external factors that influence rural tourism development in this area, from which strengths, weaknesses, opportunities, and threats (SWOT) analysis was conducted. This study suggests that leadership commitment is necessary to drive the process of developing rural tourism through capability building.

Keywords

Rural tourism · Rural tourism resources · Rural tourism capabilities · Rural tourism strategies · Rural tourism activities

6.1 Introduction

Researchers (Mwesiumo et al., 2022; Wijijayanti et al., 2020) view rural tourism in various ways; however, there are some key elements that all academics share: first, rural tourism is based in rural spaces distinct from urban environments; second, it has the distinctive cultural characteristics of rural areas; and third, it is closely integrated with the rural traditional primary industry (Chi et al., 2020; Dašić et al., 2020; Dinis et al., 2019). Rural tourism can help rural communities achieve Sustainable Development Goals (SDGs) by increasing their economic, social, and environmental well-being (Lara-Morales & Clarke, 2022; Marini Govigli et al., 2022). Rural tourism is also perceived by the 2030 Agenda for Sustainable Development (United Nations, 2022)

N. Z. Mgabhi (✉) · N. S. Mzimela
Faculty of Humanities and Social Sciences, Department of Recreation and Tourism, University of Zululand, KwaDlangezwa, South Africa

I. O. Ezeuduji
Department of Recreation and Tourism, University of Zululand, KwaDlangezwa Campus, Zululand, South Africa
e-mail: ezeudujii@unizulu.ac.za

K. Dube et al. (eds.), *Tourism in Rural Protected Areas*, Sustainable Development Goals Series,
https://doi.org/10.1007/978-3-031-87333-1_6

as an auxiliary activity to support sustainable development goals (SDGs).

Rural tourism has the potential to help address SDG's Goal 1—"End poverty in all its forms everywhere." It may aid people in rural communities by providing employment opportunities and sources of income, assisting in poverty reduction, and fostering economic development (Birendra et al., 2021). Rural tourism development can also aid in achieving SDG's Goal 8—"decent work and economic growth." It can encourage entrepreneurship, create jobs, and foster economic growth in rural areas (Winter & Kim, 2021). According to Wijijayanti et al. (2020), rural tourism destinations contribute to community empowerment and long-term economic development. Therefore, capability development for rural tourism is a long-term, holistic approach to empowering rural communities to take advantage of the potential of tourism for economic growth and collaborative well-being while protecting their cultural and natural resources (Setokoe & Ramukumba, 2022). Capability development is the process of establishing new competencies or improving current ones (Teece, 2016). According to Bakhru and Gleadle (2010) and Bakhru et al. (2013), capability denotes the extent to which resources are combined and coordinated. Ezeuduji (2015a, 2015b) asserts that rural tourism development is an entrepreneurial process that requires rural communities to highlight the capabilities and knowledge necessary to embark on tourism development.

The subject of rural tourism has been thoroughly studied (An & Alarcon, 2020; Kumar & Shekhar., 2020; Sipatau et al., 2020). However, there is a knowledge gap and a dearth of literature on the importance and effectiveness of capability development in rural tourism (Kumar & Shekhar., 2020). Therefore, this study focuses on rural tourism resources and capability development by investigating local communities' strengths, weaknesses, opportunities, and threats (SWOT) to rural tourism. SWOT analysis is a useful tool in rural tourism research (Li, 2022; Zhao et al., 2021). SWOT analysis has been widely used to help researchers, policymakers, and stakeholders understand the rural tourism context, make informed decisions, and create strategies that build on opportunities, address weaknesses, embrace strengths, and reduce threats (Li, 2022; Saikim et al., 2020). As much as resources and capability development may be situated within the rural tourism development's internal environment, the external environment impacts this development. This may facilitate or deter rural tourism. Hence, SWOT analysis (which has both internal and external perspectives) can inform this study and provide a better strategic direction.

This study was conducted with the local communities around Hluhluwe iMfolozi Park (HiP) in KwaZulu-Natal, South Africa, and is of significance in guiding similar local communities in developing nations, especially those living around a protected area on how to develop rural tourism when suitable, for their socioeconomic benefits. Trang et al. (2021) advocated that one of the challenges to the success of rural tourism development is that those involved need to gain the fundamental knowledge and abilities needed to manage and serve tourists. Hence, training and education for rural populations to work in the tourism industry is part of the development of rural tourism (Nomnian et al., 2020).

6.2 Theoretical Underpinning

Streimikiene and Bilan (2015) stated that numerous ideas on the growth of rural tourism stem from the forces of supply and demand in the market. Consumer motivation is the primary driver of the demand for the growth of rural tourism. The primary supply variable for rural tourism development is the use of an evolutionary strategy in conjunction with local tourism resources and capabilities. This is also consistent with the writings of other authors, including McComb et al. (2017), Ezeuduji (2015a, 2015b), Ezeuduji (2017), and Ezeuduji and Rid (2011).

According to Streimikiene and Bilan (2015), an evolutionary approach to tourism development holds that the growth of rural tourism is a process. As such, it should be viewed as a slow, organic process of change. This process supports a deeper understanding of the dynamism of rural tourism development. Therefore, it is crucial to determine the primary elements (which Ezeuduji, 2017 emphasized) that can impact the kind of changes in rural tourism growth as well as the best approaches for identifying and assessing them. The success of this study depends on theoretical knowledge.

6.3 Literature Review

Prior survey data from the neighborhoods surrounding HiP show that most families depend on state social subsidies to put food on the table (Nsukwini, 2015) and that there is a significant unemployment rate in the park's surrounding areas (Nsukwini & Bob, 2016). Therefore, increasing rural tourism in these areas could be one of the best ways to reduce poverty and create new jobs (Amir et al., 2015). Trukhachev (2015) asserted that rural towns, however, need to have specific resources available to them to support their endeavors to create and grow tourism infrastructure. Owing to resources such as the natural landscape, local cultures and customs, and the chance to engage in outdoor activities and consume local food, certain rural places are quite touristic (desirable to tourists).

Nsukwini and Bob (2016) asserted that it is critical to pinpoint the resources that attract visitors to rural areas and persuade them to spend money on activities or experience a rural lifestyle. Some academics, such as Zhou (2018), contend that rural tourism encompasses a wide range of activities that highlight various resources and are frequently cultural and natural. In general, rural tourism is driven mostly by people and depends on the natural and cultural resources of a unique area (Sanagustin-Fons et al., 2018).

According to Zhou (2018), rural tourism has the potential to stimulate a diverse range of innovative economic ventures, partnerships, and associations that supplement conventional activities such as farming. However, rural tourism is more than just a village vacation; it encompasses all other activities in rural areas. According to a study conducted in rural Gambia by Ezeuduji and Rid (2011), the cultural attractions of rural tourism include traditional dances, regional cuisine, music, rural lifestyles, crafts, local languages, festivals, and rituals. Because every country tailors its rural tourist development policy to its unique circumstances, rural tourism operations and amenities differ from country to country (Temelkov & Gulev, 2019). Consequently, residents can provide a wide range of activities as a component of their rural tourism. The analysis of the literature reveals that for rural communities to grow tourism, they must demonstrate the skills and knowledge they possess (Ezeuduji, 2015a; Nsukwini & Bob, 2016; Temelkov & Gulev, 2019). Hence, Ezeuduji (2017) highlights the need to acquire the three cardinal requirements of distinct value propositions, operational capabilities, and key customer relations for rural tourism development.

Tourism supports long-term development in rural areas (Nooripoor et al., 2021; Yang et al., 2021). Given the rising level of unemployment in South Africa, Nicolaides (2020) argued that the rural tourism sector is a crucial area to consider. This sector will play a significant role in the government's strategy for developing tourism if an effective and efficient environmental management framework is in place (Nicolaides, 2020). However, rural areas are often characterized by low productivity, poor education, and low incomes (Mgabhi & Ezeuduji, 2023; Woyo & Musavengane, 2023). Therefore, resources and capability development are important issues to consider while growing rural tourism, even though these topics have been ignored in many studies (Nicolaides, 2020; Yang et al., 2021).

Studies on tourism development have assumed that capability building is vital for communities, particularly those living in rural areas (Imbaya et al., 2019; Mohd Noh et al., 2020). Mgabhi and Ezeuduji (2023) claim that inadequate financial resources, a lack of skilled labor, and a lack of education and awareness are constant challenges facing rural tourism businesses.

According to Ezeuduji (2017), local stakeholders may significantly manage the development of rural tourism depending on their understanding of their intended roles (see Table 6.1).

Table 6.1 Stakeholders' demands and roles in rural tourism development

Major stakeholder group	Demands	Roles
Local communities	Income from tourism	Staging events
Tourists	What they pay for	Service consumers
State governments	Compliance to policy	Policy formulation
Local governments	Rural development	Monitoring and evaluation
National environment agencies	Environmental standards	Environment impact assessment
Tourism authorities	Popularity of destination	Marketing communications
Not-for-profit organizations	Equity, local empowerment	Initial funding, research and capability building
Tour operators	Returns on their investment	Tourists' acquisition

Source: Adapted from Ezeuduji (2017, p. 950)

According to Ezeuduji (2015a, 2015b), promoting rural tourism is a business process that calls for rural communities to showcase their knowledge and capabilities, which are essential for the development of tourism. Sanagustin-Fons et al. (2018) state that enabling elements such as the necessary knowledge and experience, awareness, information accessibility, and financial assistance determine community members' ability to participate in tourism activities. The major supply components of rural tourism development include local tourism resources and capabilities as well as the use of an evolutionary approach in rural tourism development (Ezeuduji & Mgabhi, 2023).

Therefore, this study explored the necessary resources and capability requirements for rural tourism development. With this essential knowledge and a combination of skills, the sustainability of rural tourism development is improved (Rosalina et al., 2021). It is only sufficient to have rural tourism resources in place, with the required capabilities to combine and coordinate them to produce sustainable rural tourism products and services (Arintoko et al., 2020). The research design is described in the next section.

6.4 Research Design and Methods

Hluhluwe iMfolozi Park (HiP) is known as the Hluhluwe-Umfolozi Game Reserve. This park was named Africa's oldest reserve. According to estimates, it is the same size as the Netherlands Antilles, at 96,000 hectares, or 960 km^2. The park was formally inaugurated in 1895 (Hluhluwe Game Reserve, 2018). Important tourist destinations such as the World Heritage Site-iSimangaliso Wetland Park and St. Lucia, which have many lodging options and tourist activities, are close by.

This study was conducted within local communities near a protected area (Hluhluwe iMfolozi Park (HiP)). However, data were collected from park managers, business owners/managers, and local communities to inform the study objectives from both internal and external perspectives. Tourism is important in this study area, as both domestic and international tourists visit parks frequently. However, local communities must derive more benefits from tourism activities in their area.

This study, being exploratory and interpretive in nature, used a qualitative research approach to collect data. This study used a non-probability sampling technique in the form of a purposive approach (as the study population cannot be easily estimated) to select 44 participants who were knowledgeable about the study area and could provide the information needed for the study.

Interview guides were used, and semi-structured interviews and focus group discussions were conducted with participants. This provides insight into the perceptions of park managers and local business owners/managers toward rural tourism development and what local communities need in terms of capability development to fully support rural tourism development in their area. Variables from prior studies (such as Ezeuduji & Rid, 2011; Ezeuduji, 2015a, 2015b;

Rangus et al., 2018) and the researchers' knowledge were used to guide data collection during focus groups and interviews.

Twenty semi-structured interviews with management staff from HiP and business owners/managers were conducted, and three focus groups with eight participants each were held in 2022. Men, women, and youth were separated into three different groups for the focus groups because, in African culture, the youth tend to be less vocal when surrounded by more senior citizens. Women also typically prefer men to take the lead in discussions because they are seen as heads of family (Ezeuduji & Rid, 2011). As these participants felt free to respond to questions and participate in discussions without hindrance in their groups, subdividing them according to age and gender groups created a way to extract as much information as possible. Four (4) young men and four young women were recruited to achieve gender equality within the youth group.

The study assumed that the participation of 7 park managers, 13 local business owners/managers (of tourism-related businesses such as tour operators and accommodation facilities) in individual interviews, and 24 members of the community in focus group discussions produced reliable data to address the research aim, as data saturation was reached at the third focus group, 17 park manager, and 13 business manager interviews.

The verbatim transcripts of the interviews and focus group recordings were manually subjected to data analysis using the content analysis approach. Data generated from the interviews and focus groups were not substantial and required careful interpretation, without bias, from the local knowledge point of view, as one of the researchers is a resident in the study area and understands the local ways of expression. Data were coded to provide themes and subthemes that were ultimately connected to the study's aim. Participants' opinions were organized into three categories for discussion: "*very commonly recorded*" (up to five times), "*commonly recorded*" (three or four times), and "*not commonly recorded*" not commonly recorded (one or two times). The next section presents the results and discussion of this study.

6.5 Results and Discussion

The demographic characteristics of the 44 participants are presented in Table 6.2. Those who participated in this study were recruited from the HiP management staff, tourism business owners/managers around the park, and members of the three communities neighboring the park (men, women, and youth). The men and women who participated in the study were older than the age of thirty-five (35+), and the participating youths were 18 to 35 years old. Table 6.2 illustrates the demographic characteristics of the seven HiP management staff and supervisors who agreed to participate in this study and the 13 tourism business owners and managers adjacent to the park. According to Table 6.2, business managers were more willing to participate in the research than business owners. The demographic characteristics of the 24 community members who participated in this study are presented in Table 6.2. The three focus groups comprised eight men, eight women, and eight youths aged between 18 and 55 years.

The next section presents the results of this study based on the themes and subthemes developed from the research regarding this study's goal.

6.5.1 Theme One: Rural Tourism Resources

This subsection presents the results for the rural tourism resources found and those that need to be improved in this study area. The results revealed that both managers of HiP and owners/managers of tourism businesses around the park perceived the area as one of the resource-richest areas in terms of natural and cultural resources (*commonly recorded*).

> Most people come here to experience nature (nature reserves), the big five and beautiful accommodation establishments (our chalets, which are built in African style, make our tourists feel culturally involved). (Male, HiP Manager).

> There is a variety of wild animals here which other countries or places do not have. This is a place to be or a place to know if you are not a South African or even if you are from South Africa but from other provinces. (Female, Tourism Business owner).

Table 6.2 Participants' demographics

Participant number	Gender	Age range	
Participant's occupation: HiP managers ($N = 7$)			
01	Female	25–30	
02	Female	40–45	
03	Male	30–35	
04	Male	25–30	
05	Female	40–45	
06	Male	45–50	
07	Male	35-40	
Note: All were HiP managers			
Participant's occupation: tourism business manager/owner ($N = 13$)			
08	Female	45–50	TBM
09	Female	35–40	TBM
10	Female	45–50	TBM
11	Female	50–55	TBO
12	Female	40–45	TBO
13	Female	35–40	TBM
14	Male	50–55	TBO
15	Male	35–40	TBM
16	Male	35–40	TBM
17	Male	45–50	TBO
18	Male	35–40	TBM
19	Male	30–35	TBM
20	Male	30–35	TBM
Note: TBM tourism business manager, TBO tourism business owner			
Participants: community members ($N = 24$)			
01	Men	36–55	8 members
02	Women	36–55	8 members
03	Youth	18–35	8 members
Note: All were community members			

Source: Authors' compilation

This assertion is in line with the findings of Lane and Kastenholz (2015), who stated that rural tourism uses natural, cultural, historical, and other resources as well as the characteristics of rural settlements to develop complex tourism products. Rich natural, cultural, and historical resources can be found in the South African province of KwaZulu-Natal, where the study area is situated, serving as a foundation for tourism development. From the interviews and focus groups, the subthemes emanating from the main theme (rural tourism resources) were discussed.

6.5.1.1 Subtheme 1: Financial Resources

The first subtheme concerns funds, handling finances, and financial challenges faced by rural tourism businesses. According to HiP managers, the park has meant raising money to help local communities launch their own companies such as cooperatives and other community-based initiatives *(commonly recorded)*. See statements below.

> We have something we call a community levy; a certain percentage of the money generated from entrance fees goes to local communities to start their businesses or whatever projects they want to be engaged in. (Female, HiP Manager).

> Governments do not provide financial assistance, particularly to start-up businesses; I raised capital on my own and applied for bank loans several times without success. (Male, business owner)

Focus groups were conducted with men, women, and youth in separate sessions. Individuals reported needing help when attempting to obtain funding to establish, expand, or maintain their businesses. They mention a high unemployment rate in the area. They recommended that, instead of searching for work elsewhere, they could start their businesses if they could receive financial assistance (*very commonly recorded*). This finding supports the assertion made by Radović et al. (2020) that a lack of funding limits the development of sustainable rural tourism. Therefore, the information provided by the park managers, which suggests that the park offers financial help to the people, conflicts with this, as local community members never admitted this.

> Most of our children do not work; maybe if there can be some sort of financial support presented to us, they can be able to create their employment and be able to put food on the table. (Male, community member).

> Even if we want to start tourism businesses, it is not easy because we do not have money to start, and the level of unemployment in this community is too high, more especially unemployed youth. (Youth, community member).

The findings suggest that financial resources might be beneficial for promoting rural tourism development in the HiP and surrounding areas. According to a study conducted in Serbia by Radović et al. (2020), the main reason for the lack of rural tourism development in most rural areas is the need for clear strategic development orientations and financial investments. However, the mismanagement of corporate funds by owners or members of management can be a major issue that leads to inactivity or slowed development. Local communities require training, and both semi-skilled and low-skilled businesses want relevant knowledge, communication, and market access to exhibit their products and services. Capability development is an important aspect of the study area.

6.5.1.2 Subtheme 2: Human Resource Development

This subtheme covered education and training, knowledge, and awareness among local communities. The focus group discussions revealed that educational training is important, given that community members need to understand tourism in Gunjaneni, Machibini, and eMadwaleni. Both local tourism business owners/managers and HiP staff members agreed that a significant amount of education and training in tourism-related subjects would be required for locals to acquire the necessary knowledge, which may transform into an interest in tourism *(very commonly recorded).*

> Education and training can improve local level of interest in tourism, but to some of them because it all starts with passion first; people need to be passionate about engaging in businesses. You can't just force education on them. (Female, HiP Manager).

> Education and training can assist them to be the ones that are creating employment instead of them looking for an employment from the park; they should be able to create their own facilities that will be sort of direct competition to the park itself. (Female, Business owner).

Pimonratanakan and Pooripakdee (2017) found that training is a crucial activity that can promote learning and enhance understanding and proficiency in job-related tasks. People from the local community expressed their dissatisfaction with not receiving sufficient training in tourism education and their willingness to grab the opportunity if given the chance (very commonly recorded).

> We see those big cars coming in and out of the park, they say 'abavakashi' I think that is all tourism is about, people coming from other places to watch wild animals inside the park. (Male, Community member).

> Eeeh! My child I do not know what it is except that I sometimes hear children at home talking about tourism, but I have no idea of what is it all about. (Female, Community member).

Given that people from these communities do not know much about tourism, this study indicates that the most significant capital formation in this study area's rural tourism development is developing people through education and training. The following subsection highlights the findings from HiP managers, owners of tourism businesses, and local communities regarding physical resources.

6.5.1.3 Subtheme 3: Physical Resources (Infrastructure)

This subtheme covers aspects connected to physical resources such as accommodations and roads (accessibility). The research area required more important infrastructure, such as roads and recreational facilities, which may have served as the foundation for rural tourism services *(commonly recorded)*. See the participants' responses below.

> In terms of physical resources, I can say that our area is not easily accessible due to poor quality of roads; most of our guests used to complain about it. (Female, business manager).

> This area is very beautiful in terms of natural and cultural resources, but it still lacks sub-attractions that will attract more tourists to visit the area; these can include centers for recreational activities and more accommodation establishments. (Male, HiP manager).

The responses suggest that the area has several natural beauties, but a lack of lodging options and poor road conditions may pose challenges for rural tourism development. According to Jovanović and Ivana (2016), tourism infrastructure comprises several services required to meet the needs of tourists and enhance satisfaction throughout their stay at the destination. Communities neighboring HiPs complained about inadequate accessibility and poor road conditions (*commonly recorded*). According to the information provided, the study area needs to improve infrastructure development for rural tourism development to be successful.

> We lack access to basic infrastructure; for instance, our roads consist predominantly of gravel, [and] they often deteriorate rapidly, especially in the wet season, disrupting transport services and access to health centers, towns, schools, and markets when it is most needed. (Female, community member).

> Our roads are very bad, even the one that goes inside the park is not well maintained and most of the roads are not tared. (Youth, community member).

Accessibility is an essential component of tourism. According to Dalimunthe et al. (2020), tourism development is closely tied to, and dependent on, the development of available infrastructure. The development of the existing infrastructure must aid in the development of the tourism industry.

6.5.1.4 Subtheme 4: Intangible Resources (Water and Sanitation)

Sanitation and water are major issues in the HiP area (*commonly recorded*). Most park managers expressed concerns about the lack of clean water in the area. If these important needs for tourism development are met, the brand image of a tourism destination will be positively affected (Ezeuduji & Nkosi, 2017). Some managers responded as follows.

> Our wastewater treatment infrastructure remains inadequate, water that we have is only for bathing, toilets and cleaning consumption, for clean drinking water one need to go to the shops to buy. (Female, business manager).

> We pump water from the river (Imfolozi River) using a generator; during dry seasons, we find it very difficult to supply the whole park (Male, HiP manager).

The issue of inadequate sanitation and water supply by community members also brought up the people living in the communities surrounding the HiP experience.

> We do not have water taps or tanks in the community, only rely on water from the dams and river; our local municipality does nothing about this issue; only once in a while (3 months) [do] they get to send water tanks to deliver water to the community. (Female, community member).

According to the opinions, one factor that should be considered to help the tourism industry, especially in developing nations, is access to basic sanitary facilities (Elysia & Wihadanto, 2020). A summary of the knowledge and skills related to rural tourism that the study participants possessed is provided in the following subsection.

6.5.2 Theme Two: Rural Tourism Knowledge and Capabilities

According to Ezeuduji (2015a, 2015b), promoting rural tourism is a business process that calls on rural communities to showcase their knowledge and abilities, which are essential for the development of tourism. Communities surrounding HiP should strive to acquire the requisite knowledge and skills to prepare themselves better, as the industry is always in need of fresh and updated knowledge to keep up with growing tourism trends.

6.5.2.1 Subtheme 1: Lack of Business Education

In South Africa, the tourism industry is frequently promoted to increase employment. However, there are still issues to be addressed regarding the types of jobs in the tourism industry, education levels, and skill development within the field (Booyens, 2020). Participants responded:

> Most locals lack knowledge on how to operate businesses successfully, which is a result of not having a proper education. This makes it difficult for a community member to start a business and to be able to manage it on their own successfully. (Male, Business manager).

> People from those communities are not well educated on tourism and other business operations, although there is youth that is now starting to pop up. We are forced to employ people coming from faraway places, mostly in management positions. (Male, HiP Manager).

These assertions indicate that educating locals about business operations is something that needs to be done *(very commonly recorded).* Ilban and Kaşlı (2015) assert that education plays a critical role in driving economic growth by creating a skilled labor force that industries require and by increasing worker productivity. Community members unanimously agreed that they possessed the necessary qualities to successfully operate and manage businesses on their own *(very commonly recorded).*

> Although there are no formal business qualifications that we hold, generally, I think we can be able to run and manage a business successfully. (Male, community member).

> I would say that we, as black people, are not into tourism because if you look at the tourism businesses found here are dominantly owned by whites. (Youth, community member).

According to this study, governments must educate local communities about resources for starting and operating businesses. Youths should also be awarded bursaries to enable them to pursue education, as most participants noted that they lacked official business qualifications.

6.5.2.2 Subtheme 2: Lack of Trust for Cooperation

The results revealed a lack of collaboration and trust among community members, which contributed to the downfall of several cooperatives in the area *(commonly recorded).*

> There is some sort of a backpacker business by the gate that was funded by [the] government; it was cooperatively owned by residents, but it is no longer operating, and now no one wants to be blamed for its failure. (Female, HiP manager).

Some participants indicated a lack of trust in cooperative businesses, while others were still interested in becoming involved in cooperatives *(commonly recorded).*

> We once tried a cooperative group whereby we were constructing bricks, the group failed, [and] finances disappeared into thin air; I do not know if I can be able to be involved in any cooperatives program. (Youth, community member).

> I have heard of some cooperative businesses that went wrong, but that does not mean we are no longer interested in it; if the opportunity can present itself, I will certainly jump to it. (Male, community member).

According to Qurniati et al. (2017), enhancing social capital and raising productivity requires [s] several elements including trust, cooperation, and teamwork. The results of the study show that locals prefer to work alone because they do not trust each other.

6.5.2.3 Subtheme 3: Lack of Tourism Business Experience

The poor management of rural tourism businesses may result from inexperience. Residents in the park's surrounding areas needed more operational expertise and understanding in the field of tourism. According to the findings, communities lack an appropriate tourism experience that would have enabled them to become involved in rural tourism development *(very commonly recorded).*

> Local communities do not have people who have operated tourism businesses or maybe who have experienced working in any tourism venture. The park provides training for them because we have lodges here in the park whereby we strictly employ residents (Female, HiP manager).

> We do have locals working here, but mostly in [the] maintenance department because I find it hard to hire them, [the] reason being, they do not have required job experience and qualifications. (Male, business manager).

It was noted from the focus groups that most participants needed more business or work experience linked to tourism. Despite their lack of expertise, they stated that they could work in cleaning, waiting, and running their businesses with the right guidance *(Very commonly recorded).*

> I do not have enough work experience, but I am very good with people, and I know that being in tourism requires one to be friendly to customers. (Youth, community member).

> Cleaning is part of our everyday chores; that way, I can be able to work in [the] housekeeping department only if I can secure employment. (Female, community member).

According to these findings, locals have little expertise in identifying commercial, marketing, and tourism-related opportunities (Rasoolimanesh et al., 2018). Having the right abilities and qualities is crucial for successful and productive engagement in the tourism industry. The next subsection discusses rural tourism strategies.

6.5.3 Theme 3: Rural Tourism Strategies

Based on the findings, theme 3 offers targeted recommendations for developing rural tourism in the areas surrounding the HiP. The subtheme that follows discusses training and development as a strategy that could be key to the development of rural tourism in the study area.

6.5.3.1 Subtheme 1: Training and Development

The findings demonstrated that among the strategies used to encourage rural tourism development, the focus should be on training and development *(most commonly recorded).*

> I think that series of training and educational courses should be offered to encourage new entrants to tourism and to help existing businesses. (Female, business owner).

> Training can go a long way [in] helping us to improve the little knowledge that we have about tourism and in assisting us to welcome tourism development positively. (Youth, community member).

The results of the study by Purnomo et al. (2020) indicate that developing human resources and educating them is one strategy that should be used to promote tourism and increase the role that communities play in managing tourism potential. This study suggests that residents should regularly receive education and training to increase their knowledge, awareness, and skills.

6.5.3.2 Subtheme 2: Rural Tourism Cooperatives

The establishment of rural cooperatives was noted as a potentially excellent means of fostering communities and engaging locals in potential rural tourism activities *(very commonly recorded).* Ezeuduji (2017) also noted change management in sub-Saharan Africa's development of rural tourism.

> Cooperatives can play a major self-help role in the areas surrounding HiP, particularly because they could provide opportunities for employment, education, potable water, improved sanitation, roads, and market access to the residents. It is easy to voice out opinions and grievances if you are in groups that when the one is standing alone. (Female, business owner).

Local communities also agreed that they would seize the possibility of participating in rural tourism cooperatives if supported.

> Rural tourism cooperatives can help us eradicate hunger and reduce the level of unemployment we are faced with in this community. (Male, community member).

In the areas surrounding HiP, cooperatives might be considered one of the approaches to actively engage in rural tourism. To guarantee that cooperatives have access to the people and leadership abilities required to conduct their business in the long run, this study suggests capability development for members, which may include ways of fostering trust among themselves.

6.5.3.3 Subtheme 3: Business Plan in Place

For residents to attract financial interest, one of the HiP managers emphasized the significance of creating a business plan *(not commonly recorded).*

> If residents can put together a detailed business plan to submit to potential funding structures, an effort could be made because really, no one would just throw their money away without any assurance of profitable and long-term investment in black and white. (Male, HiP manager).

The findings of this research indicate that residents should draft business plans to plan better, guide, and make critical decisions that will increase their chances of receiving financing and, ultimately, increase their chances of success.

6.5.4 Theme four: Rural Tourism Activities

The activities that the local communities can offer tourists are outlined in this section.

6.5.4.1 Subtheme 1: Cultural Activities

Most participants mentioned cultural dancing as the most popular cultural practice in the communities surrounding the HiP *(Most commonly recorded).* This suggests that areas like HiP and nearby communities may profit from cultural activities by expanding the selection of cultural heritage experiences offered to visitors, as some tourists may be interested in getting a taste of the "local" and "everyday" culture of the host community. In a discussion of strategic event-based rural tourism development in sub-Saharan Africa, cultural experience was regarded as an essential offering to rural tourists (Ezeuduji, 2015b).

> Residents perform [the] cultural dance (ingoma) for guests, to entertain them and teach them how Zulus dance, after performing they get tips/ money from guests as a sign of appreciation (Female, HiP manager).

> We have cultural dance groups here in this community, but they only used to perform in traditional gatherings for free; there is nothing they get out of it. (Youth, community member).

According to this study, cultural dances reflect the customs and way of life of locals and can be utilized as a rural tourism activity to attract tourists.

6.5.4.2 Subtheme 2: Arts and Craft

According to the findings of a study conducted by Nyawo and Mubangizi (2015), the arts and crafts sector is critical for rural communities since it addresses some of the socioeconomic difficulties that communities face and contributes significantly to tourism. Many people in communities surrounding HiP rely on the production and sale of traditional arts and crafts as a source of income *(very commonly recorded).* Participants responded:

> Most people from these communities are very talented in handicrafts products, and most of them depend on it to put food on the table (Female, HiP manager).
> This community is very rich when it comes to the production of arts and crafts products. You will find men selling wood trays by the roads and women selling grass mats and other handicraft products to earn their living (Male, business manager).

Communities near HiPs feature a wide range of arts and crafts products made by residents (very commonly recorded).

> I specialize in making ethnic costumes (amabheshu), I also make wood products that include trays and spoons. I also use those ethnic costumes to hire it to people (Male, community member).
> As women in this community, we make grass mats and sell them to people who are coming from Durban; they take them on low prices and make big money out of it there (Female, community member).

New markets for local artists can be created through rural tourism. For example, the park may purchase local artwork to decorate its rooms and lobby areas and employ local artists, dancers, and singers to entertain guests.

6.6 Capability Development for Rural Tourism: SWOT Analysis

SWOT analysis is a useful tool for assessing the feasibility and potential success of operationalizing capability development for rural tourism. In operationalising capability development for rural tourism, it is essential to leverage strengths and opportunities while actively addressing weaknesses and mitigating threats (Fridayani & Chiang, 2022). The assessment of strengths and weaknesses is illustrated in Fig. 6.1, which also looks for opportunities and threats by examining how the external environment affects the host communities.

Addressing weaknesses in the development of rural tourism capabilities is crucial for the industry's long-term viability and sustainability. Operationalization strategies used to address the key weaknesses in the study area are presented in Table 6.3.

Internal

Positive	**Strengths** ✓ Artistic dance as a tourist attraction. ✓ The presence of various arts and crafts activities. ✓ Untouched natural beauty, rich wildlife species, and major cultural and historical features. ✓ The young population with knowledge of tourism and a positive attitude toward rural tourism development, and the elderly population with Indigenous knowledge about the area. ✓ Friendliness of the rural population.	**Weaknesses** ✓ Lack of skilled and trained human capital. ✓ Residents' unfamiliarity with and lack of training about tourism. ✓ Lack of local government's support in funding local communities. ✓ Lack of trust between community members. ✓ Lack of tourism infrastructure. ✓ No existing clear functional framework for rural tourism development. ✓ There is no adequate interpretation of all aspects of the region's natural, cultural, and historical heritage.	**Negative**
	Opportunities ✓ Likelihood of funding from local government. ✓ Sunny climate in the region. ✓ Emerging niche market focusing on individual travel. ✓ The Western perception that Africa has the most authentic rural areas. ✓ The region is not sensitive to dangerous natural disasters. ✓ Proximity to HiP. ✓ No recorded significant negative socio-cultural and environmental impact from tourism.	**Threats** ✓ Economic, commercial, and logistical issues such as accessibility and infrastructure to the area, availability of skills, and interest of investors. ✓ Growing competition within the global tourism industry, especially with the emergence of virtual travel and new information and communication technology.	

External

Fig. 6.1 Capability development: SWOT analysis (Authors' compilation)

Table 6.3 Operationalization strategies

Key weaknesses	Operationalization strategies
Lack of skilled and trained human capital in the area	– Encouraging local tourism businesses to provide on-the-job training opportunities for community members – Partnering with local schools, colleges, and training centers to design courses and programs tailored to the needs of the rural tourism industry
Residents' unfamiliarity with and lack of training about tourism	– Providing businesses with training on how to handle changes in the industry – Providing training and support to locals to enhance their service capabilities
Lack of local government's support in funding local communities to make their own living through tourism	– Working with local officials to simplify permitting and licensing processes – Developing public–private partnerships to access government support
Lack of trust between community members	– Developing clear benefit-sharing agreements to ensure that the economic gains from tourism are distributed fairly
Lack of tourism infrastructure	– Investing in improving roads, bridges, and transportation networks – Collaborating with nearby transportation providers to improve accessibility
No existing clear functional framework for rural tourism development	– Forming a working group by involving every relevant rural tourism stakeholder – Review existing laws, regulations, and policies related to tourism and identify areas that need revision or adaptation to support rural tourism
There needs to be an adequate interpretation of all aspects of the region's natural, cultural, and historical heritage	– Creating interpretive materials, such as brochures, signage, and audio guides, that offer educational content about the natural, cultural, and historical heritage – Offer guided tours led by trained interpreters who can provide in-depth knowledge and storytelling to visitors

Source: Authors' compilation

Table 6.3 provides how rural tourism can overcome accessibility challenges, limited services, and environmental concerns through strategic planning, infrastructure development, and capability building. Furthermore, trust must be established among community members, and a defined functional framework for development can drive the industry's growth responsibly and inclusively. Adequate interpretation of natural, cultural, and historical resources can enhance visitor experiences while preserving a region's unique identity.

6.7 Theoretical and Practical Implications

This study adds to the body of knowledge on rural tourism by offering an assessment of its potential, strengths, weaknesses, opportunities, and threats. Researchers also investigated the different resources that the study area had to offer. Understanding these resources could help to formulate long-term rural tourism development initiatives. The study provides insight into the development of policies that could facilitate investment in rural tourism infrastructure, encourage capability building programs, and create a framework that contributes to rural tourism entrepreneurship. All of this supports and strengthens the evolutionary approach to rural tourism development.

Beneficial developments can be seen in rural areas' infrastructure, cooperative formation, training center establishment, and partnerships with educational institutions and other rural tourism stakeholders. Organizing local events and rural exhibitions where residents can demonstrate their talents to tourists, as well as introducing new types of rural tourism activities, may also aid in the development of successful rural tourism.

6.8 Conclusions and Recommendations

Developing rural tourism in communities is an alternative strategy to strengthen the local economy. The problem lies in rural communities' ability to implement and manage the creation of a critical mass of products and services associated with tourism. The study's findings revealed that locals need more specialized training and experience to start operating a business. Therefore, the creation of local education and training programs is crucial for the effective planning of rural tourism in the study area. This supports the evolutionary approach to rural tourism development, which is at the core of the theoretical base of this study. Implementing capability development for rural tourism is a complicated and dynamic process that necessitates a comprehensive approach, active collaboration, and adaptability. Rural communities can reach their full potential as dynamic and sustainable tourism destinations by carefully addressing their weaknesses and threats.

Based on the study findings and the theoretical approach consulted (evolutionary theory), the study proposes the following recommendations.

1. To enhance rural tourism offerings, it is important to include other local attractions, including locally made handicrafts, in the product mix.
2. It is recommended to hold small local markets regularly to sell handicrafts and other products manufactured by locals.
3. Leadership commitment is required to drive the process of developing rural tourism through capability building, which includes educating locals, helping them create business strategies, and securing funding to launch their businesses.
4. Training and experience are required for several positions in the tourism industry. Internships, industry-sponsored apprenticeship programs, on-the-job training, and other diverse forms of education and training programs are a few solutions that the study area may take into consideration.

It is possible to argue that rural tourism development could be a viable solution for renewing the area's features and activities (village renewal), making it more appealing to tourists, and incorporating rural tourism into the social, cultural, and economic aspects of HiP's surrounding communities. This can aid in the creation of jobs, the development of small businesses, and the promotion of local innovation. The study contends that if the recommendations mentioned above are thoroughly adopted, rural tourism in this area can be replanned, effectively implemented, and sustained.

6.9 Limitations and Areas for Future Studies

This study can be conducted in a wider context, at the provincial or national level, to generate more robust findings that will support the creation of a provincial or national rural tourism development framework. More research should be conducted to investigate the long-term effects of capability building programs on rural tourism, particularly the long-term viability of businesses and communities. Rural tourism research can also be undertaken on the successful techniques of knowledge transfer and training that empower local communities and businesses.

References

Amir, A. F., Abd Ghapar, A., Jamal, S. A., & Ahmad, K. N. (2015). Sustainable tourism development: A study on community resilience for rural tourism in Malaysia. *Procedia-Social and Behavioral Sciences, 168*, 116–122. https://doi.org/10.1016/j.sbspro.2014.10.217

An, W., & Alarcon, S. (2020). How can rural tourism be sustainable? A systematic review. *Sustainability, 12*(18), 7758. https://doi.org/10.3390/su12187758

Arintoko, A., Ahmad, A. A., Gunawan, D. S., & Supadi, S. (2020). Community-based tourism village development strategies: A case of Borobudur tourism village area, Indonesia. *Geo Journal of Tourism and Geosites, 29*(2), 398–413. https://doi.org/10.30892/gtg.29202-477

Bakhru, A., & Gleadle, P. (2010). *Competing with capabilities*. The Open University Business School.

Bakhru, K. M., Sanghi, T., Seema, M., & Medury, Y. (2013). A principal component analysis of teaching competencies required for management education. *Arth Prabhand: A Journal of Economics and Management, 2*(7), 23–29. https://studylib.net/doc/8102555/a-principal-component-analysis-of-teaching-competencies-r

Birendra, K. C., Dhungana, A., & Dangi, T. B. (2021). Tourism and the sustainable development goals: Stakeholders' perspectives from Nepal. *Tourism Management Perspectives, 38*, 100822. https://doi.org/10.1016/j.tmp.2021.100822

Booyens, I. (2020). Education and skills in tourism: Implications for youth employment in South Africa. *Development Southern Africa, 37*(5), 825–839. https://doi.org/10.1080/0376835X.2020.1725447

Chi, X., Lee, S. K., Ahn, Y. J., & Kiatkawsin, K. (2020). Tourist-perceived quality and loyalty intentions towards rural tourism in China. *Sustainability, 12*(9), 1–18. https://doi.org/10.3390/su12093614

Dalimunthe, D. Y., Valeriani, D., Hartini, F., & Wardhani, R. S. (2020). The readiness of supporting infrastructure for tourism destination in achieving sustainable tourism development. *Society, 8*(1), 217–233. https://doi.org/10.33019/society.v8i1.149

Dašić, D., Živković, D., & Vujić, T. (2020). Rural tourism in the development function of rural areas in Serbia. *Ekonomika Poljoprivrede, 67*(3), 719–733. https://doi.org/10.5937/ekoPolj2003719D

Dinis, I., Simões, O., Cruz, C., & Teodoro, A. (2019). Understanding the impact of intentions in the adoption of local development practices by rural tourism hosts in Portugal. *Journal of Rural Studies, 72*, 92–103. https://doi.org/10.1016/j.jrurstud.2019.10.002

Elysia, V., & Wihadanto, A. (2020). The impact of poor sanitation on tourism development: A global review. *Indonesia Journal of Urban and Environmental Technology, 3*(2), 220–231. https://doi.org/10.25105/urbanenvirotech.v3i2.6720

Ezeuduji, I. O. (2015a). Building capabilities for sub-Saharan Africa's rural tourism services performance. *African Journal for Physical Health Education, Recreation and Dance, 21*(Supplement 2), 68–75. https://doi.org/10.10520/EJC182967

Ezeuduji, I. O. (2015b). Strategic event-based rural tourism development for sub-Saharan Africa. *Current Issues in Tourism, 18*(3), 212–228. https://doi.org/10.1080/13683500.2013.787049

Ezeuduji, I. O. (2017). Change management for sub-Saharan Africa's rural tourism development. *Current Issues in Tourism, 20*(9), 946–959. https://doi.org/10.1080/13683500.2014.946892

Ezeuduji, I. O., & Mgabhi, N. Z. (2023). Potential for rural tourism development around Hluhluwe iMfolozi Park, South Africa: A local community perspective. *African Journal of Development Studies, 13*(1), 89–113. https://doi.org/10.31920/2634-3649/2023/v13n1a5

Ezeuduji, I. O., & Nkosi, G. S. (2017). Tourism destination competitiveness using brand essence: Incorporating the 'zuluness' of the Zulu Kingdom. *African Journal of Hospitality, Tourism and Leisure, 6*(1), 1–10.

Ezeuduji, I. O., & Rid, W. (2011). Rural tourism offer and local community participation in The Gambia. *Tourismos: An International Multi-disciplinary, 6*(2), 187–211. https://doi.org/10.26215/tourismos.v6i2.227

Fridayani, H. D., & Chiang, L.-C. (2022). Local governments' strategy on the sustainability of micro enterprises amidst the uncertainty of the COVID-19 pandemic: SWOT analysis. *Local Administration Journal, 15*(2), 119–138. https://so04.tci-thaijo.org/index.php/colakkujournals/article/view/256470

Hluhluwe Game Reserve. (2018). *Hluhluwe game reserve - Hluhluwe Imfolozi Park*. Retrieved May 4, 2024, from https://hluhluwegamereserve.com/hluhluwe-imfolozi-park-2

Ilban, M. O., & Kaşlı, M. (2015). The impact of tourism education problems on the students' affective commitment. *European Journal of Social Sciences, 46*(4), 390–403.

Imbaya, B. O., Nthiga, R. W., Sitati, N. W., & Lenaiyasa, P. (2019). Capacity building for inclusive growth in community-based tourism initiatives in Kenya. *Tourism Management Perspectives, 30*, 11–18. https://doi.org/10.1016/j.tmp.2019.01.003

Jovanović, S., & Ivana, I. L. I. Ć. (2016). Infrastructure is an important determinant of tourism development in the countries of Southeast Europe. *Ecoforum Journal, 5*(1), 288–294. https://ideas.repec.org/a/scm/ecofrm/v5y2016i1p34.html

Kumar, S., & Shekhar. (2020). Technology and innovation: Changing concept of rural tourism–A systematic review. *Open Geosciences, 12*(1), 737–752. https://doi.org/10.1515/geo-2020-0183

Lane, B., & Kastenholz, E. (2015). Rural tourism: The evolution of practice and research approaches towards a new generation concept. *Journal of Sustainable Tourism, 23*(8-9), 1133–1156.

Lara-Morales, O., & Clarke, A. (2022). Sustainable tourism value chain analysis as a tool to evaluate tourism's contribution to the sustainable development goals and local Indigenous communities. *Journal of Ecotourism, 23*(2), 129–147. https://doi.org/10.1080/14724049.2022.2147189

Li, K. (2022). SWOT analysis of e-commerce development of rural tourism farmers' professional cooperatives in the era of big data. *IET Communications, 16*(5), 592–603. https://doi.org/10.1049/cmu2.12358

Marini Govigli, V., Rois-Díaz, M., den Herder, M., Bryce, R., Tuomasjukka, D., & Górriz-Mifsud, E. (2022). The green side of social innovation: Using sustainable development goals to classify environmental impacts of rural grassroots initiatives. *Environmental Policy and Governance, 32*(6), 459–477. https://doi.org/10.1002/eet.2019

McComb, E. J., Boyd, S., & Boluk, K. (2017). Stakeholder collaboration: A means to the success of rural tourism destinations? A critical evaluation of the existence of stakeholder collaboration within the Mournes, Northern Ireland. *Tourism and Hospitality Research, 17*(3), 286–297. https://doi.org/10.1177/1467358415583738

Mgabhi, N. Z., & Ezeuduji, I. O. (2023). Stakeholder perceptions of the potential for rural tourism development around Hluhluwe iMfolozi Park in South Africa. *Revista Turismo & Desenvolvimento, 42*, 45–65. https://doi.org/10.34624/rtd.v42i0.32664

Mohd Noh, A. N., Razzaq, A. R. A., Mustafa, M. Z., Nordin, M. N., & Ibrahim, B. (2020). Elements of community capacity building (CCB) for cbet development. *PalArch's Journal of Archaeology of Egypt, 17*(9), 4970–4981. https://archives.palarch.nl/index.php/jae/article/view/4731/4666

Mwesiumo, D., Halfdanarson, J., & Shlopak, M. (2022). Navigating the early stages of a large sustainability-oriented rural tourism development project: Lessons from Træna, Norway. *Tourism Management, 89*, 104456. https://doi.org/10.1016/j.tourman.2021.104456

Nicolaides, A. (2020). Sustainable ethical tourism (SET) and rural community involvement. *African Journal of Hospitality, Tourism and Leisure, 9*(1), 1–16. https://www.ajhtl.com/uploads/7/1/6/3/7163688/article_1_vol_9_1__2020_unisa.pdf

Nomnian, S., Trupp, A., Niyomthong, W., Tangcharoensathaporn, P., & Charoenkongka, A. (2020). Language and community-based tourism use, needs, dependency, and limitations. *Austrian Journal of South-East Asian Studies, 13*(1), 57–79. http://eprints.sunway.edu.my/id/eprint/1326

Nooripoor, M., Khosrowjerdi, M., Rastegari, H., Sharifi, Z., & Bijani, M. (2021). The role of tourism in rural development: Evidence from Iran. *GeoJournal, 86*(4), 1705–1719. https://doi.org/10.1007/s10708-020-10153-z

Nsukwini, S., & Bob, U. (2016). The socio-economic impacts of ecotourism in rural areas: A case study of Nompondo and the Hluhluwe-iMfolozi Park (HiP). *African Journal of Hospitality, Tourism and Leisure, 5*, 1–15. http://www.ajhtl.com/uploads/7/1/6/3/7163688/article_14_vol_5__3__final.pdf

Nsukwini, S. B. (2015). An investigation of the socio-economic impacts of ecotourism in rural areas: A case study of Nompondo, a community bordering the Hluhluwe-iMfolozi Park (HiP), KwaZulu-Natal, South Africa. In *A doctoral thesis submitted to the University of KwaZulu-Natal, South Africa*. University of KwaZulu-Natal.

Nyawo, J., & Mubangizi, B. C. (2015). Art and craft in local economic development: Tourism possibilities in Mtubatuba local municipality. *African Journal of Hospitality, Tourism and Leisure, 4*(2), 1–15.

Pimonratanakan, S., & Pooripakdee, S. (2017). The human resource development in the learning organization for the organizational development. *International Journal of Business and Economic Affairs, 2*(3), 183–192. https://doi.org/10.24088/IJBEA-2017-23003

Purnomo, S., Rahayu, E. S., Riani, A. L., Suminah, S., & Udin, U. (2020). Empowerment model for sustainable tourism village in an emerging country. *The Journal of Asian Finance, Economics and Business, 7*(2), 261–270. https://doi.org/10.13106/jafeb.2020.vol7.no2.261

Qurniati, R., Febryano, I. G., & Zulfiani, D. (2017). How does trust influence social capital to support collective action in agroforestry development? *Biodiversitas Journal of Biological Diversity, 18*(3), 1201–1206. https://doi.org/10.13057/biodiv/d1844

Radović, G., Petrović, M. D., Demirović Bajrami, D., Radovanović, M., & Vuković, N. (2020). Can proper funding enhance sustainable tourism in rural settings? Evidence from a developing country. *Sustainability, 12*(18), 1–16. https://doi.org/10.3390/su12187797

Rangus, M., Brumen, B., & Topler, J. P. (2018). Sustainable tourism development in rural areas: The role of stakeholders. *Academica Turistica-Tourism and Innovation Journal, 10*(2), 167–173. https://doi.org/10.26493/2335-4194.10.167-173

Rasoolimanesh, S. M., Jaafar, M., & Tangit, T. M. (2018). Community involvement in rural tourism: A case of Kinabalu National Park, Malaysia. *Anatolia, 29*(3), 337–350. https://doi.org/10.1080/13032917.2017.1412327

Rosalina, P. D., Dupre, K., & Wang, Y. (2021). Rural tourism: A systematic literature review on definitions and challenges. *Journal of Hospitality and Tourism Management, 47*, 134–149. https://doi.org/10.1016/j.jhtm.2021.03.001

Saikim, F. H., Peters, R. F., Simon, N., Angki, T. A., & Bagul, A. H. B. P. (2020). Rural tourism destination strategy through SWOT analysis in Kg. Kolosunan, Penampang, Sabah. *Journal of Contemporary Issues and Thought, 10*, 36–45. https://doi.org/10.37134/jcit.vol10.4.2020

Sanagustin-Fons, V., Lafita-Cortés, T., & Moseñe, J. A. (2018). Social perception of rural tourism impact: A case study. *Sustainability, 10*(2), 1–25. https://doi.org/10.1080/02642069.2019.1570152

Setokoe, T. J., & Ramukumba, T. (2022). Ability to do for self. A rural community's view of the importance of capacity building in sustainable tourism development. *Geo Journal of Tourism and Geosites, 41*(2), 583–588. https://doi.org/10.30892/gtg.41232-866

Sipatau, J. A., Mapjabil, J., & Imang, U. (2020). Diversity of rural tourism products and challenges of rural tourism development in Kota Marudu, Sabah. *Journal of Islamic, Social, Economics and Development, 5*, 51–61. http://www.jised.com/PDF/JISED-2020-34-12-06.pdf

Streimikiene, D., & Bilan, Y. (2015). Review of rural tourism development theories. *Transformations in Business and Economics, 14*(2), 21–34. http://www.transformations.knf.vu.lt/35/ge35.pdf

Teece, D. J. (2016). Capability development. In M. Augier & D. Teece (Eds.), *The Palgrave encyclopedia of strategic management*. Palgrave Macmillan. https://doi.org/10.1057/978-1-349-94848-2_572-1

Temelkov, Z., & Gulev, G. (2019). Role of crowdfunding platforms in rural tourism development. *SocioBrains, 56*, 73–79. https://eprints.ugd.edu.mk/id/eprint/22063

Trang, N. T. H., Trang, N. T. T., & Lam, N. T. (2021). Developing rural tourism with the participation of related parties in Bac Kan Province. *Journal La Bisecoman, 2*(1), 13–22. https://doi.org/10.32628/GISRRJ21325

Trukhachev, A. (2015). Methodology for evaluating the rural tourism potentials: A tool to ensure sustainable development of rural settlements. *Sustainability, 7*(3), 3052–3070. https://doi.org/10.3390/su7033052

United Nations. (2022). *The sustainable development goals report 2022*. Retrieved September 29, 2023, from https://unstats.un.org/sdgs/report/2022/

Wijijayanti, T., Agustina, Y., Winarno, A., Istanti, L. N., & Dharma, B. A. (2020). Rural tourism: A local economic development. *Australasian Accounting, Business and Finance Journal, 14*(1), 5–13. https://doi.org/10.14453/aabfj.v14i1.2

Winter, T., & Kim, S. (2021). Exploring the relationship between tourism and poverty using the capability approach. *Journal of Sustainable Tourism, 29*(10), 1655–1673. https://doi.org/10.1080/09669582.2020.1865385

Woyo, E., & Musavengane, R. (2023). Has community-based tourism been a useful tool for poverty eradication in Southern African rural destinations? *African Journal of Hospitality, Tourism and Leisure, 12*(2), 521–534.

Yang, J., Yang, R., Chen, M. H., Su, C. H. J., Zhi, Y., & Xi, J. (2021). Effects of rural revitalization on rural tourism. *Journal of Hospitality and Tourism Management, 47*, 35–45. https://doi.org/10.1016/j.jhtm.2021.02.008

Zhao, M., Zhou, J., & Mu, J. (2021). SWOT research on the development of rural tourism E-commerce system under the background of big data era. *Mobile Information Systems, 2021*, 1–13. https://doi.org/10.1155/2021/8112747

Zhou, J. (2018). Status, causes and countermeasures of environmental pollution in China's rural tourism development. *Nature, Environment and Pollution Technology, 17*(2), 543–549. https://neptjournal.com/upload-images/NL-64-29-(27)G-143.pdf

Part III

Critical Debates for Tourism in Protected Areas in Southern Africa

7 The Efficacy of Visitor Management Tools in Encouraging Ethical Practices: Evidence from iSimangaliso Wetland Park, South Africa

M. Kuseni, Uwe P. Hermann, and P. P. S. Sifolo

Abstract

This chapter aims to assess the efficacy of visitor management tools in encouraging ethical practice at iSimangaliso Wetland Park (IWP). This study is divided into two parts. The first part consists of a literature review. The literature review identifies and adopts five visitor management tools: infrastructure designs and eco-labels, education and awareness programs, codes of conduct, economic instruments and regulatory systems, and social norms. In the second part, a survey of 323 respondents at IWP has been used to test the efficacy of the visitor management tools. Data obtained is then analysed using descriptive (frequency) and inferential statistics (ANOVA). The study results revealed variations in the efficacy of visitor management tools in encouraging visitors to engage in ethical practice at IWP. Infrastructure designs, eco-labels, and education and awareness programs emerged as the most effective tools for promoting ethical practice at IWP from a visitor's point of view. This research will help visitor management in the future. By understanding visitor management tools' efficacy, IWP authorities can select and fund the most effective tools for encouraging visitors to engage in ethical practice. This study is the first to assess the efficacy of infrastructure designs and eco-labels, education and awareness programs, codes of conduct, economic instruments and regulatory systems, and social norms at IWP from the visitors' point; hence, it contributes to visitor management literature.

Keywords

ISimangaliso Wetland Park · Visitor management · Visitor management tools · World heritage site

7.1 Introduction

The global tourism market has been growing exponentially in the post-COVID-19 period, recouping an estimated monetary value of $631 billion in 2022, which is a growth of 45% from the pandemic period (Dube & Nhamo, 2024; Globaldata, 2022). The report suggests that about 900 million visitors traveled internationally in

M. Kuseni (✉) · U. P. Hermann · P. P. S. Sifolo
Department of Tourism Management, Centre for Sustainable Tourism, Tshwane University of Technology, Pretoria, Republic of South Africa
e-mail: kuseniM@tut.ac.za; hermannup@tut.ac.za; sifoloPS@tut.ac.za

K. Dube et al. (eds.), *Tourism in Rural Protected Areas*, Sustainable Development Goals Series,
https://doi.org/10.1007/978-3-031-87333-1_7

2022 (IBISWorld, 2024; United Nations World Tourism [UNWTO], 2020). The market size is expected to grow by a further 3.47% between 2024 and 2028, resulting in a market volume of US$1063.00 billion in 2028 (IBISWorld, 2024). The South African tourism market has been experiencing a similar trajectory, recording approximately 7.6 million international visitors between January and December 2023, a remarkable 51% increase compared to the same period in 2022.

With such a rapid increase in tourist numbers globally and in South Africa in particular, managers of protected areas such as the iSimangaliso Wetland Park (IWP) face challenges in achieving their dual mandate of preserving sensitive natural resources and recreation provision (Ervinal et al., 2020; Eyassu et al., 2021). Large visitors to whom park managers are trying to offer unique areas for creating memorable experiences are leaving imprints through littering, vegetation trampling and degrading in situ natural and cultural resources (Agnew & Delmas, 2013; Nqalamba et al., 2023), and IWP may not be sidestepped. If left unmanaged, visitors' negative impacts may affect sensitive tourism resources, which may lead to a decline in visitor satisfaction. Consequently, protected area managers face challenges in implementing effective visitor management tools to enhance resource sustainability (Dissanayake & Arachchi, 2018).

Visitor management tools are implemented to control visitor numbers, improve their experiences, create enjoyment, and modify their on-site behavior (Eyassu et al., 2021). In this regard, park managers regulate visitor use through zoning, seasonal pricing to control the type of visitation, differentiated entry or user fees, restricting access to only accredited organizations, providing interpretation programs, and the use of codes of conduct. Park managers may also resort to creating safety barriers and information centers to encourage visitors to engage in ethical practice. Although these tools exist, questions regarding their appropriateness, application, and efficacy in different contexts remain. This view is supported by Albrecht (2017) who noted that the efficiency of visitor-management tools depends on the size, application, regulations, and political parameters that exist in different contexts.

Studies have been conducted based on different visitor management tools in different contexts. For example, Kamau et al. (2020) conducted a study at the Mount Kenya World Heritage Site assessing the efficacy of codes of conduct in promoting responsible tourism practices. In this study, visitor codes of conduct were found to contribute effectively toward improved responsible behavior. Juma et al. (2020) conducted their study in the Massai Mara Nature Reserve which revealed nature interpretation and awareness as effective tools for unlocking the full potential of responsible behavior among visitors. Dolnicar et al. (2020) studied the influence of rewards on pro-environmental behavior and found that rewards may positively influence pro-environmental behavior. Gössling et al. (2019) proved the effectiveness of social norms on responsible behavior, while León and Araña (2020) justified raising awareness as a useful visitor management tool. Alemu and Assefa (2024) and Nqalamba et al. (2023) investigated several strategies such as zoning, site hardening, site discrimination, and waste management systems. Despite these studies confirming the efficacy of different visitor management tools, no study has yet been conducted to assess the efficacy of these tools at the IWP. Considering this background and research gap, in this chapter, we propose and analyze the efficacy of five selected visitor management tools from a visitor perspective.

7.2 Literature Review

Understanding the efficacy of visitor management tools in IWP is of utmost importance in developing existing knowledge. This section of this chapter discusses visitor management and five selected visitor management tools in light of previous studies.

7.2.1 Visitor Management

Often confused with destination management, visitor management is a summation of all practices and programs implemented to ensure visitors realize quality experiences (Bakiewicz et al., 2017; Juma et al., 2020; Nqalamba et al., 2023). According to Eyassu et al. (2021), visitor management mitigates potential negative visitor impacts while simultaneously enhancing visitors' comfort and safety by minimizing potential hazards and risks. Effective visitor management in any area is important because it has the potential to profoundly shape positive visitor experiences. However, not all visitors perceive visitor management positively as some feel that it denies them the freedom to create memorable experiences. Despite visitor management being criticized, Bhati and Pearce (2017) believe that it is a necessary evil that imposes restrictions on visitors' freedom while preserving their safety and welfare.

Visitor management in protected may be implemented in two, that is, manipulating the demand side and the supply side of tourism services. On the demand side, Alemu and Assefa (2024) suggested four scenarios that protected area managers should adopt: (a) controlling visitor numbers, (b) understanding visitor practices, (c) improving visitor experience and creating enjoyment at the destination, and (d) modifying visitors' on-site practice. On the supply side, Manning et al. (2017) suggested four scenarios for protected area managers: (a) increasing the supply, (a) reducing the impact of use, (c) increasing the durability of the resource experience, and (d) limiting the use of resources. To increase the supply, the managers may increase the size of the area to accommodate larger visitor numbers or open extra hours to shape the flow of tourists while reducing the impact of use. Managers do not eliminate activities that damage the environment but alter their timing, location, or practices. To increase the durability of the resource, protected area managers create "hard surfaces," such as walkways, to absorb the direct physical impacts of visitor activities. Finally, by limiting the use of resources, managers limit problematic visitors or ban visitor uses that might promote problems or challenges.

7.2.2 Visitor Management Tools

Protected area managers use different management tools to mitigate the negative impacts of visitors. Eyassu et al. (2021) and Nqalamba et al. (2023) categorize contemporary visitor management tools into two major groups: hard and soft visitor management tools. Hard visitor management tools mostly rely on restrictions and punitive measures to manipulate positive human practice (Alvarez-Sousa & Prados, 2020). Price discrimination, building animal viewpoints and pathways, erection of fences, site hardening, zoning, closure, penalties, patrols, or limiting access are some of these potential visitor management tools (Leask & Wanhill, 2022). Hard visitor management tools may apply punitive measures to visitors who engage in inappropriate behavior. Considering their punitive nature, hard visitor management tools are primarily suitable for short-term use (Juma et al., 2020).

On the other hand, soft visitor management tools are subtle enforcement of rules to guide visitors, using tools such as signage, interpretative messages, and codes of practice, as well as developing barriers to restricting certain activities (Nqalamba et al., 2023). These visitor management tools are generally non-obstructive (Alemu & Assefa, 2024) because they inform and persuade visitors to behave appropriately using signage and maps (Ervinal et al., 2020). While categorizing visitor management tools may be justified as it enhances understanding, one should note that applying them in isolation might only realize short-term outcomes (Juma et al., 2020). For example, soft visitor management tools such as education and awareness may be applied to complement hard visitor management tools such as zone closure. In this section of the chapter, we discuss five visitor management tools to formulate propositions.

7.2.2.1 Visitor Codes of Conduct

Visitor codes of conduct are the most common strategies used to manage visitors in protected areas. Visitor codes of conduct refer to prescribed actions articulated as either a social norm or rule of law sanctioned by enforcement authorities (Lovelock & Lovelock, 2013). The codes of conduct for visitors serve as a set of guidelines or rules that all visitors must follow. In the tourism context, Kamau et al. (2020) refer to codes of conduct as rules, instructions, and obligations that determine how visitors should behave in a specific area. Visitor codes of conduct educate visitors to reduce their negative impacts on sensitive tourism resources and conscientise them not to engage in practices that may be deemed provocative by host community members. The influence of codes of conduct in promoting responsible practice has been proven by Kamau et al. (2020); however, their influence relies on the ability of people to interpret them (Delistavrou et al., 2017). Thus, there is a need to monitor and evaluate their efficacy in different contexts (Marion & Reid, 2007) such as IWP. This will provide empirical evidence of whether it is effective to implement it.

7.2.2.2 Economic and Regulatory Instrument

Protected area authorities may use economic and regulatory instruments to manipulate positive ethical practices. Economic measures denote the use of strategies such as price discrimination, fines, and fees to regulate visitor flows. Price discrimination involves setting different prices for the same recreational opportunity based on whether visitors are local or international, peak season or low season and location (Alemu & Assefa, 2024). Prices may be differentiated based on the sensitivity of the area to visitor numbers (Leung et al., 2018). Authorities may permit visitors to participate in supervised or restricted activities. While economic and regulatory systems are perceived as achieving better results than visitor codes of conduct (Pearce, 2013), they inherit negative connotations from visitors because of their emphasis on restricted actions. Marion and Reid (2007) observe that imposing regulations might antagonize visitors rather than win their support. Furthermore, enforcing the law and imposing penalties might incur expenses for park authorities as more manpower is required.

7.2.2.3 Social Norms

Another visitor management tool common in visitor management literature is social norms which refer to a collective understanding of behavioral expectations of society (Dolnicar et al., 2020) or reactive actions by a group of people against an inappropriate act (Pearce, 2013). Despite social norms being underutilized within the ethical practice domain, it is clear that within collectivistic societies, the goal of a group of people plays a vital role in changing behavior. Actions of a group of people may have the potential to exert social pressure which may force people to change their behavior (Kim & Seock, 2019; Reynolds et al., 2015). In the tourism context, social norms may act as inter-visitor behavioral controls, where statements are being made to discourage others from engaging in inappropriate behavior in tourism attractions. However, it should be noted that mechanisms such as inter-visitor behavior control need to be implemented with a consensus because they might promote or create conflict among visitors (Pearce, 2013).

7.2.2.4 Education and Awareness Programs

Another visitor management tool useful to protected area managers is the provision of education and awareness programs. Also known as interpretation, education, and awareness programs are soft and non-obstructive on-site visitor management tools implemented to encourage the conservation and improvement of attractive areas (Ballantyne et al., 2021; Kamau et al., 2020). Education and awareness programs build a deeper appreciation for the areas visited and empower visitors to make sustainable choices. More importantly, education and awareness programs link an area's resources to visitors.

Signboards and brochures can used to educate and create visitor awareness (León & Araña, 2020). Lu (2013) reported that well-informed visitors treat their holiday experiences and settings more sympathetically than those with poor expectations of what a setting can offer. In addition, Juvan and Dolnicar (2014) note that providing information and guidance about what visitors might see and experience can reduce stress and promote mindful, self-aware, and responsible behavior.

7.2.2.5 Infrastructure and Eco-labels

Destination managers may encourage stakeholders to alter their physical infrastructure to become more environmentally friendly to encourage visitors to ethically practice. Some managers use eco-labels to promote pro-environmental behaviors such as reusing, recycling, refusing, recognizing, reducing, replacing, and re-engineering (Dolnicar et al., 2020). In many destinations, green hotels are increasingly being constructed and given eco-label certification to encourage ethical practices (Dolnicar et al., 2019). With eco-label certification, service providers are given a written guarantee that their business services conform to specific requirements. Budeanu (2007) observed that eco-labels may be useful instruments with highly informative values that influence visitor attitudes. Nowadays, eco-labels are emerging as the most important instrument for informing tourists to use green products/services.

Based on the above discussion, it is evident that the five identified visitor management tools (codes of conduct, economic and regulatory instruments, social norms, education and awareness, infrastructure design, and eco-labels) are important in influencing positive visitor behavior in protected areas. However, it is unclear how these visitor management tools are implemented in different contexts and how visitors' subsequent attitudes affect their efficacy (Juma et al., 2020). Juma et al. (2020) further noted that the situation is further complicated, as some protected areas have limited resources to implement, monitor, and evaluate visitor management tools. Thus, it is important to conduct empirical studies in nature-based areas, such as IWP, which thrives on the uniqueness of tourism resources, to assess the efficacy of different management tools from a visitor's point of view.

7.3 Methodology

The methodology of this study will be described in this section. Firstly, the section provides a brief overview of the study area. This is followed by the research design and approach and a description of how the data is analysed. Lastly, an outline of ethical considerations is given.

7.3.1 Study Context

The study was conducted in the context of the IWP, a wetland park located in the KwaZulu-Natal province, South Africa. The park has natural landscapes, flora, and fauna. Combining 16 parcels of previously fragmented pieces of land, the park stretches approximately 87 km from Kosi Bay in the north to Maphelane in the south (iSimangaliso Wetland Park Authority [IWPA], 2018). Touted as a flagship of the South African post-apartheid protected area conservation policy (Hansen, 2014), the park hosts some of the best-selling tourism features in the province. As such, visitor numbers are expected to increase substantially. Increased visitors to the park are desirable; however, unpleasant visitor behavior and practices such as littering, feeding wild animals, moving close to wild animals, making noise, and vandalizing vegetation bring the necessity of regulating visitor activities. To avert inappropriate visitor behavior that might lead to extreme environmental decay and monumental epidemic loss, park authorities must manage visitors in a sustainable way to balance the increased visitation and sustainability of tourism resources.

7.3.2 Research Design and Approach

The study adopted a descriptive research design using a survey to collect data from visitors to the IWP. The study population comprised 522,954 visitors to the park (iSimangaliso Wetland Park Authority [IWPA], 2018). The Raosoft sample size calculator was employed to calculate the sample size considering a confidence limit of 5% and a 95% confidence interval resulting in a sample size of 323 respondents. Convenient and purposive sampling techniques were used to access the samples. Data were collected between November 1 and December 30, 2021, and these months were considered appropriate based on their peak visitation periods. The survey consisted mainly of closed items rated on a five-point Likert scale. The researchers themselves distributed the instruments to visitors. Respondents were given the instrument while entering the park and asked to drop it at the main ports of exit or the reception of the accommodation that they had used.

The instrument was developed based on the literature on visitor management strategies (Alemu & Assefa, 2024; Ballantyne et al., 2021; Juma et al., 2020; Kamau et al., 2020; Pearce, 2013). The instrument consists of two sections. In the first section, respondents were asked to provide their social demographics (e.g., gender, age, and education), while the second section investigated the respondents' perceptions of the efficacy of five selected visitor management tools in changing their behavior using a 5-point Likert scale (1 = extremely effective, 2 = neutral, 3 = partially effective, 4 = effective, and 5 = very effective). Cronbach's alpha was used to assess the internal consistency (internal reliability) of the test items. Cronbach's alpha values greater than 0.7 are considered acceptable, while values less than 0.5 are unacceptable. In this study, when the items in the research instrument were tested for reliability, the lowest alpha score was 0.6673, the highest was 0.6827, and the average score was 0.7388, indicating reliability.

7.3.3 Data Analysis

Quantitative data was collated and analysed using SPSS V27 software for descriptive and inferential statistics to answer the study objective. A two-step data analysis was performed: first, the means and standard deviations of the visitor management tools were calculated; second, an ANOVA analysis was carried out. Bartlett's equal variance test was performed to identify statistically significant differences between the demographic groups in terms of visitor management tools.

7.3.4 Ethical Consideration

Ethical aspects of confidentiality, privacy, informed consent, and voluntary participation were addressed as respondents were not asked to provide their names or contact details (Bell, 2014). The researcher assured the respondents that the study was for academic purposes only and that the research would not cause them any harm. The study was ethically approved at the Tshwane University Of Technology Research Ethics Committee with the ethics number FCRE2021/FR/07/001-MS(2). Permission to conduct the research using participants was also requested and given by the IWP authority before the study was undertaken.

7.4 Results Presentation and Discussion

7.4.1 Demographic Profile of the Respondents

The study achieved a sample (n) size of 323 respondents, which accounted for an 85% response rate after administering 380 questionnaires in total. Of the 323 respondents, 62% were domestic tourists, and 38% were international tourists. Fifty percent of the respondents were male, 49% were female, and 1% had never indicated their gender. A total of 23% of the sample were aged between 18 and 30, 35% were aged between 31 and 40, and another 31% were aged between 41 and 50. These results show a generally young sample, with 89% aged below 50 years and a small fraction above 51. This result could be attributed to the fact that the IWP is only accessed through the road, which is strenuous and offers more adventurous activities, which may be more challenging for senior citizens. Most respondents were educated, with 39% being matriculants, 22% being 3-year diploma/degree holders, and 20% being 4-year degree holders. Such an educated population suggests that the IWP is a destination that offers many activities for educational tourists. A total of 37% of the sample were married, 27% lived with partners, and 26% were single. Most respondents were vacationing at the IWP with their family members (50%), spouses/partners (17%), or alone (16%). Most respondents were repeat visitors (84%) who paid to enter the park (93%); 29% of the respondents had visited the park twice, while 29% had visited once.

7.4.2 Perceived Efficacy of Visitor Management Tools

Respondents were asked to rate the efficacy of the proposed visitor-management tools related to their compliance. The descriptive findings are presented in Table 7.1.

The findings showed that four of the five selected visitor management tools were more effective as they were rated positively by the respondents. Out of the four perceived as the more effective visitor management tools by respondents, infrastructure designs and eco-labels were considered the most effective visitor management tools, with a mean score of 4.16. The ratings were skewed toward the upper end of the scale, indicating a generally positive perception. The second-highest-rated visitor management tool was education and awareness, with a mean of 4.06. Visitor codes of conduct and economic and regulatory systems were found in average positions, with codes of conduct rated with a mean score of 3.84, while economic and regulatory systems had a mean score of 3.79. The importance of these four visitor management tools with high effectiveness in influencing visitors to comply with ethical practices is supported by various scholars in the literature (e.g. Alemu & Assefa, 2024; Ballantyne et al., 2021; Juma et al., 2020; Kamau et al., 2020). These findings suggest that, as currently implemented in the IWP, infrastructure designs and eco-labels, education and awareness, codes of conduct, and economic and regulatory systems may be considered effective visitor management tools for promoting ethical practice. On the other hand, the results suggest that social norms may not be effective in promoting responsible behavior at IWP as the efficacy was rated low. Social norms do not seem appropriately practiced at IWP; hence, they have the lowest mean score of 2.89. Table 7.2 summarizes the varying degrees of effectiveness of each factor in encouraging ethical visitor behavior at IWP.

7.4.3 Comparison of Visitor Management Tools Based on Demographic Groups

The next stage of data analysis involved determining whether any significant relationships existed between the efficacy of the visitor management tools and socio-demographic characteristics of the sample, including gender (female, male, and other), age (18–30, 31–40, 41–50, 51–60, 60+), education (below grade 12, 3-year diploma/degree, 4-year degree, postgraduate), marital status (single, living with partner, married, divorced, widow/widower, and other), paying options (non-paying, paying, and others), accompany (alone, family member, and friend), and frequency of visit (once, twice, 3 times, 4 times, 5 times, and over). ANOVA was conducted, and the results are presented in Table 7.3.

No significant differences in respondents' sentiments toward the efficacy of visitor management tools in IWP were found concerning gender ($p = 0.0718$) or different age groups ($p = 0.1631$). However, significant differences were found in

Table 7.1 Perceived efficacy of visitor management tools

Visitor management tools	1	2	3	4	5	Mean	Standard deviation	Cronbach alpha
	Response as percentage							
Visitor codes of conduct	1.8	5.9	24.5	41.8	26.0	3.84	0.940	0.6827
Economic instruments and regulatory systems	0.3	8.7	28.2	36.8	26.0	3.79	0.936	0.6673
Social norms, e.g., inter-person behavior	19.5	24.8	16.4	25.7	13.6	2.89	1.350	0.7694
Education and awareness, e.g., guided tours	0.9	4.4	17.0	42.4	35.3	4.06	0.882	0.6796
Infrastructure designs and eco-labels	0.7	4.3	18.9	30.3	45.8	4.16	0.922	0.6754

Source: Authors' compilation

Table 7.2 Visitor management tool kit for efficacy in encouraging ethical behavior

Visitor management tools	Visitor management tools components	Example	Effectiveness in encouraging ethical visitor behavior
Infrastructure design and eco-labels	Infrastructure design	Eco-friendly accommodations, waste management systems, renewable energy sources	High
	Eco-labels	Green certifications and eco-friendly labels for products/services	High
Education and awareness	Education programs	Workshops, seminars, educational materials	High
	Awareness campaigns	Public service announcements, social media campaigns	High
Economic instruments and regulatory systems	Economic incentives	Tax incentives for eco-friendly practices, subsidies for sustainable businesses	High
	Regulatory systems	Environmental regulations, zoning laws	High
Visitor codes of conduct	Codes of conduct	Leave no trace principles, ethical wildlife viewing guidelines	Moderate
	Social norms	Peer pressure, community expectations	Low

Source: Authors' compilation

the sentiments of different marital groups (p = 0.0011). Bartlett's equal variance test found significant differences between the ratings of visitors who identified themselves as single and those who were married (p = 0.003). Those who were married achieved a mean score of 50.916, whereas the single respondent group achieved a mean score of 12.152. Married respondents rated visitor management tools higher than single respondents. Thus, married respondents were more inclined to act ethically than single respondents.

No significant differences in respondents' sentiments regarding the efficacy of visitor management tools were found concerning educational level ($p = 0.5819$), companion ($p = 0.4439$), or paying options (p = 0.0715). Respondents' sentiments regarding visitor management tools differed concerning the frequency of visits (p = 0.0031). Repeat visitors rated the efficacy of visitor management tools more positively than first-time visitors. Repeat visitors may have a vested interest in the park, resulting in a greater inclination to act ethically.

Table 7.3 Comparison of visitor management tools based on demographic groups

Demographic and behavioral variables	Visitor management tools				
Gender	Sum of squares	df.	Mean square	*F*	Sig
Between groups	67.04187	2	33.5209375	2.66	0.0718
Within groups	4039.64209	320	12.6237565		
Total	4106.64396	322	12.7535527		
Age					
Between groups	83.1799453	4	20.7948963	1.64	0.1631
Within groups	4023.46402	318	12.6524026		
Total	4106.64396	322	12.7535527		
Marital status					
Between groups	254.58199	5	50.916398	4.19	0.0011
Within groups	3852.06197	317	12.1516151		
Total	4106.64396	322	12.7535527		
Educational level					
Between groups	36.6293227	4	9.15733068	0.72	0.5819
Within groups	4070.01464	318	12.7987882		
Total	4106.64396	322	12.7535527		
Frequency of visits					
Between groups	204.145979	4	51.0364947	4.10	0.0031
Within groups	3299.17254	265	12.4497077		
Total	3503.31852	269	13.0234889		
Accompany					
Between groups	47.7258058	4	11.9314515	0.93	0.4439
Within groups	4058.91816	318	12.7638936		
Total	4106.64396	322	12.7535527		
Paying options					
Between groups	67.1560681	2	33.5780341	2.66	0.0715
Within groups	4039.48789	320	12.6233997		
Total	4106.64396	322	12.7535527		
Travel purpose					
Between groups	214.853651	5	42.9707301	3.50	0.0043
Within groups	3891.79031	317	12.276941		
Total	4106.64396	322	12.7535527		

Source: Authors' compilation

7.5 Conclusions

This chapter focuses on assessing the efficacy of five selected visitor management tools using the IWP as a case study. Despite focusing on the efficacy of five visitor management tools at the IWP, this chapter also contributes to the visitor management literature. A review of the literature underscores the importance of infrastructure design and eco-labels, education and awareness, economic instruments and regulatory systems, visitor codes of conduct, and social norms toward ensuring the protection of tourism resources and the provision of high-quality recreational experience in protected areas. While a considerable number of visitor management tools exist, only sporadic attention has been paid to the effectiveness of addressing and controlling the impacts of tourism (Marion & Reid, 2007). Thus, the most important contribution of this chapter is shedding light on the efficacy of the five selected visitor management tools in the context of IWP. Interestingly, a current assessment of the efficacy of five selected visitor management tools in IWP showed that some of the visitor management tools were perceived as more proficient in a good situation by visitors, whereas others were in the infant stages. The study found that infrastructure design and eco-labels, education and awareness, economic instruments and regulatory systems, and visitor codes of conduct were more effective than social norms in encouraging responsible visitor behavior. The study implies that implementing and developing better visitor codes of conduct, economic and regulatory instruments, infrastructure designs and eco-labels, and education and awareness programs at the IWP might facilitate visitors to comply with appropriate practices.

Although this study helps fill the gap in the existing knowledge in the literature and proposes some managerial implications, limitations remain. First, it utilized a convenient and purposive sample, thus limiting the generalizability of the results. Second, the results of the study were site-specific; future studies could be conducted at other sites or using different sites to make a cross-site comparison. Third, the study selected only five visitor management tools, hence ignoring a raft of visitor management tools that might have an influence, including zoning, restricted entry, and differential pricing structure. Lastly, the study used a mono-quantitative method and hence lacked mutual validation of data and findings, resulting in a less coherent and complete picture of the investigated domain (Kelle, 2006). In light of these limitations, there is a need to carry out research in future using both visitors' and destination managers' perspectives to reduce the bias of the results. Furthermore, further research into the development of visitor management tools is required to develop these tools from a constructivist point of view, which may enable greater buy-in from tourists, leading to improved ethical visitor behavior.

References

Agnew, N., & Delmas, M. (2013). Visitor management and carrying capacity at World Heritage Sites in China, case of Magao, Grottes. Extended Abstracts of the International Colloquium, Los Angeles, USA.

Albrecht, N. J. (2017). *Visitor management in tourism destination*. CABI International.

Alemu, H. T., & Assefa, Y. T. (2024). Balancing heritage preservation and tourism: Visitor management in Fasil Ghebbi, Ethiopia. *Klover Journal of Tourism and Hospitality Management, 12*(1), 19–30.

Alvarez-Sousa, A., & Prados, J. L. P. (2020). Visitor management in World Heritage Destination before and after COVID-19, Angkor. *Sustainability, 12*, 129929.

Bakiewicz, J., Leask, A., Barron, P., & Rakic, T. (2017). Using heritage interpretation to manage film-induced tourism at heritage visitor attraction. In *Visitor management in tourism destinations*. CABI.

Ballantyne, R., Hughes, K., Lee, J., Packer, J., & Sneddon, J. (2021). Facilitating zoo/aquarium visitor's adoption of environmentally sustainable behaviour. Developing a value-based interpretation. *Tourism Management, 84*, 1–11.

Bell, J. (2014). *Doing your research project: A guide for first-time researchers in social science, education and health* (4th ed.). McGraw-Hill Education.

Bhati, A., & Pearce, P. (2017). Tourist behaviour, vandalism, and stakeholder responses. In *Visitor management in tourism destination*. CABI.

Budeanu, A. (2007). Sustainable tourist behaviour: A discussion of opportunities for change. *International Journal of Consumer Studies, 31*(5), 499–508.

Delistavrou, A., Katrandjiev, H., & Tilikidou, I. (2017). Understanding ethical consumption: Types and antecedents. *Economic Alternatives, 4*, 612–632.

Dissanayake, D. M. K. D., & Arachchi, R. S. W. (2018). *Tourist attitudes of visitor management techniques in Sigiriya for the visitor experience quality*. Paper delivered at the 3rd Interdisciplinary Conference of Management Research, Sabaragamuwa University of Sri Lanka.

Dolnicar, S., Juvan, E. B., & Grun, B. (2020). Reducing the plate waste of families at hotel buffets: A quasi-experimental field study. *Tourism Management, 80*(42), 1–5.

Dolnicar, S., Knežević, C. L., & Grün, B. (2019). Changing service settings for the environment – how to reduce negative environmental impacts without sacrificing tourist satisfaction. *Annals of Tourism Research, 76*, 301–304.

Dube, K., & Nhamo, G. (2024). Tourism resilience and challenges in Limpopo, South Africa: A post-COVID-19 analysis. *Development Southern Africa, 41*, 686–703.

Ervinal, E., Wulung, S. R. P., & Octaviany, V. (2020). Tourist perception of visitor management strategy in North Banding protected area. *Journal of Business of Hospitality and Tourism, 14*(4), 340–357.

Eyassu, A. W., Asefa, G. M., & Atlug, A. (2021). Practices and challenges of visitor management implementation for sustainable tourism development in Fasil Ghebbi, Ethiopia. *Journal of Hospitality Management and Tourism, 12*(1), 1–8.

Globaldata. (2022). Retrieved March 9, 2024, from https://www.globaldata.com/data-insights-travel-tourismglobaldata

Gössling, S., Araña, J. E., & Aguiar-Quintana, T. (2019). Towel reuse in hotels: Importance of normative appeal designs. *Tourism Management, 70*, 273–283.

Hansen, M. (2014). *Struggles over conservation space: Social justice in the iSimangaliso Wetland Park*. PhD thesis. Lund, Lund University.

IBISWorld. (2024). Retrieved March 9, 2024, from https://www.ibisworld.com/global/market-research-reports-global

iSimangaliso Wetland Park Authority [IWPA]. (2018). Retrieved September 3, 2020, from https://isimangaliso.com/newsflash/isimangalisos-tree-labels

Juma, L. O., Bakos, I. M., & Khademi-Vidra, A. (2020). Nature interpretation and visitor management objective: A survey of tourists at Maasai Mara Nature Reserve, Kenta. *Sustainability, 12*, 7246.

Juvan, E., & Dolnicar, S. (2014). The attitude-behaviour gap in sustainable tourism. *Annals of Tourism Research, 48*, 76–95.

Kamau, A. M., Misiko, J., & Ndivo, R. M. (2020). Efficacy of codes of conduct in promoting responsible tourism practices within the Mount Kenya World Heritage Site. *IOSR Journal of Humanities and Social Sciences, 27*(2), 40–51.

Kelle, U. (2006). Combining qualitative and quantitative methods in research: Practices, purposes, and advantages. *Qualitative Research in Psychology, 3*(4), 293–311.

Kim, S. H., & Seock, Y. K. (2019). The roles of values and social norms on personal norms and pro-environmentally friendly apparel products purchasing behaviour. The mediating role of personal norms. *Journal of Retailing and Consumer Services, 51*, 83–90.

Leask, A., & Wanhill, S. (Eds.). (2022). *Managing visitor attractions* (2nd ed.). Elsevier.

León, C. J., & Araña, J. E. (2020). Tourist sustainable behaviour and personal communication. *Annals of Tourism Research, 85*(C), 102897.

Leung, Y., Spenceley, A., Hvenegaard, G., Buckley, R., & Groves, C. (2018). *Tourism and visitor management in protected areas: Guidelines for sustainability*. IUCN.

Lovelock, B., & Lovelock, K. (2013). *The ethics of tourism: Critical and applied perspectives*. Routledge.

Lu, H. (2013) *Learning outcomes for chinese outbound group tourists*. PhD thesis, James Cook University, Townsville, Queensland, Australia.

Manning, R. E., Anderson, L. E., & Pettengill, P. (2017). *Managing outdoor recreation: Case studies in the national parks*. CABI Publishing.

Marion, J. L., & Reid, S. E. (2007). Minimising visitor impacts to protected areas: The efficacy of low impact education programmes. *Journal of Sustainable Tourism, 15*(1), 5–27.

Nqalamba, B., Tseane-Gumbi, L., & Slabbert, E. (2023). Analysis of visitor management strategies at protected areas: The case of Pilanesberg National Park. In *ISCONTOUR 2023 tourism research perspective proceeding of the international student conference in tourism research*. Books on Demand.

Pearce, P. L. (2013). Visitor behaviour in ecotourism settings. In R. Ballantyne & J. Packer (Eds.), *International Handbook of Ecotourism* (pp. 119–134). Edward Elgar.

Reynolds, K., Subasic, E., & Tindall, K. (2015). The problem of behaviour change: From social norms to an ingroup focus. *Social and Personality Psychology Compass, 9*(1), 45–56.

United Nations World Tourism [UNWTO]. (2020). *Why tourism*? Retrieved June 5, 2021, from http://www2.unwto.org/content/whytourism

Travel Motivation Factors to Protected Areas

8

Nomfundo Sinethemba Mzimela,
Ikechukwu O. Ezeuduji,
Sibusiso David Ntshangase,
and Nelisile Zanele Mgabhi

Abstract

Protected areas worldwide help conserve biodiversity while offering opportunities for outdoor recreation and tourism. This chapter presents a recent study conducted in South Africa that explored travel motivation (push and pull) factors for tourists visiting protected areas. A structured questionnaire was used to collect data from randomly selected 435 tourists surveyed at various key tourist sites within the Hluhluwe-iMfolozi Park. The findings show that "relaxation" and "family togetherness" are essential tourist push factors, while the key tourist pull factors identified are "natural resources, destination resources, learning experience, history, and culture." Based on these findings, it is recommended that park management promote and sustain activities that address the identified key push and pull motivation drivers. Other recommendations aimed at policy formulation were organized into themes (marketing and promotion of Hluhluwe-iMfolozi Park and enhancing visitor expectations). The management focus on protected areas should be more than just transactional and selling nature. More can be done to contribute to global agreements and frameworks, such as the Sustainable Development 2030 Agenda.

Keywords

Protected areas · Nature-based tourism · Push factors · Pull factors · Visitor satisfaction

8.1 Introduction

Globally, researchers consider tourism a viable tool for boosting local economies and environmental conservation (Abrahams, 2019; Dube & Nhamo, 2020; Ramukumba, 2019). The 2030 Agenda for Sustainable Development (United Nations, 2022) perceives tourism as a supplementary activity that supports Sustainable Development Goals (SDGs). In this context, the United Nations (UN) recognition of 2017 as "the International Year of Sustainable Tourism for Development (IYSTD)" provides a unique opportunity to rethink and enhance the tourism sector's contribution to wider development (World Bank Group, 2017). Goal 8—"decent work and economic growth" suggests yearly economic growth of at least 7% per year in the least developed nations such as South Africa (SA), and

N. S. Mzimela · S. D. Ntshangase (✉) · N. Z. Mgabhi
Department of Recreation and Tourism, University of Zululand, KwaDlangezwa, South Africa
e-mail: NtshangaseSD@unizulu.ac.za

I. O. Ezeuduji
Department of Recreation and Tourism, University of Zululand, KwaDlangezwa Campus, Zululand, South Africa
e-mail: ezeudujii@unizulu.ac.za

K. Dube et al. (eds.), *Tourism in Rural Protected Areas*, Sustainable Development Goals Series,
https://doi.org/10.1007/978-3-031-87333-1_8

nature-based tourism can be a powerful driver in achieving this goal.

Protected areas (PAs) worldwide help conserve biodiversity while offering opportunities for outdoor recreation and tourism (Le Corre et al., 2021; Leung et al., 2018; Masina et al., 2021). In SA, PAs provide tourists with experiences ranging from recreational to spiritual experiences (Casteel & Bridier, 2021). However, these experiences can be influenced by specific aspects of PAs, such as their uniqueness, attractiveness, security, and management (Masina et al., 2021). The quality of the products and services provided by PAs is equally important in rescuing the SDGs and delivering meaningful experiences to tourists (United Nations, 2022). PAs are becoming increasingly concerned with balancing tourist retention (Le Corre et al., 2021) to attract new visitors (Leung et al., 2018) due to growing competition in tourism.

Travel motivational factors have a long history and continue to be a subject of discourse among scholars and policymakers even today (Masina et al., 2021; Wambani et al., 2020). Despite the growing importance of these factors in tourism, there has been a dearth of attention devoted to identifying the key "push and pull" factors and how these factors interact statistically to influence the experience of nature-based tourism, particularly in SA, using PAs as a case study. Many appeals have been made for a comprehensive understanding of the factors that impact tourist satisfaction and offer valuable data for management decision-making (Agyeman et al., 2019; Casteel & Bridier, 2021). Consequently, various studies have been conducted globally focusing on travel motivational factors (Cengizci et al., 2020; Dean & Suhartanto, 2019; Le Corre et al., 2021; Unguren et al., 2021). However, many studies that measure tourists' travel motivation factors have been widely conducted in general tourism fields (see Carvalho et al., 2023; Nordbø et al., 2023), while the "pull–push" framework is too broad to effectively measure travel motivation (Woyo & Slabbert, 2020). Hence, there is a need to identify more specific travel motivational factors for different destination contexts.

Despite extensive research conducted in the field of tourism, few studies have specifically examined travel motivations for nature-based tourism in SA. Previous research on this topic has mostly focused on developed Western countries (see Casteel & Bridier, 2021), thus leaving a theoretical gap in the understanding of African developing countries, such as SA (Woyo & Slabbert, 2020). Furthermore, existing studies on travel motivation factors in tourism have been limited in scope (Kara & Mkwizu, 2020), as they have solely concentrated on aspects of satisfaction and loyalty. Therefore, it is important to conduct continuous research on these factors, known as push and pull factors, because human behavior is subject to change over time (Woyo & Slabbert, 2020), particularly in the ever-evolving tourism industry. Moreover, a study by Woyo and Slabbert (2020) investigated the travel motivations, satisfaction and loyalty of tourists visiting a distressed destination. Woyo and Slabbert (2020) identified travel motivation dimensions and challenges using exploratory factor analysis (EFA), and further analyses were performed using correlations and regression analyses. Therefore, this study builds on the studies mentioned above by using structural equation modeling (SEM) and statistically examining the relationships between push and pull travel motivation factors.

The objective of the present study, outlined in this chapter, is to establish and statistically validate the relationships between key travel motivation factors and their interrelationships within the context of nature-based tourism experiences. This investigation was conducted in SA using Hluhluwe-iMfolozi Park (HiP) KwaZulu-Natal as a case study. This location was chosen for its tourism importance and recognition as an area that is highly specialized for nature-based tourism, where such an investigation is necessary, and motivation factors for this destination would be unique. Specifically, this study seeks to answer the following research questions: (1) What are the "push and pull factors" of key tourists when visiting PAs, such as HiP? (2) how is the relationship between "the push sub-factors" (relaxation, social interaction, family togetherness, and prestige motivations) and "pull sub-factors" (resources, learning experience, recreational facilities, natural resources, history, and culture) for tourists who visited the park, and is there a significant correlation between these two sub-dimensions? (3) What is the relationship between the composite push and pull factors for tourists who visited the park, and is there a significant correlation between these two study constructs?

The "push and pull" factors were grounded in two (2) theories: Maslow's hierarchy of needs theory (1943) and Dann's theory of push and pull motivations (1977), which are underpinned by research evidence. The study reported in this chapter is significant for enhancing the service rendered to tourists visiting PAs, using the marketing principle of identifying, understanding, and addressing the multiplicity of tourists' needs and the relationships between these needs, as this is critical for customer relationship marketing. Therefore, creating more innovative and needs-addressing packages is important for tourists. The emphasis in PAs should be more than just transactional—solely on selling nature (game drives and nature education)–as there is more that can be done. The literature relevant to this chapter is examined in the next section.

8.2 Literature Review

8.2.1 Travel Motivation and Tourist Behavior

Tourism scholars have developed an interest in travel motivation research to address the question of what motivates tourists to travel and better understand tourists' needs and behaviors (Maghrifani et al., 2021; Mapingure et al., 2019; Said & Maryono, 2018; Ying & Krishnapillai, 2018). According to the researchers, travel motivation refers to the various needs and desires that drive tourists to a specific destination. Moreover, tourist behavior relates to tourists' actual behavior and the activities they participate in while visiting a particular destination (Maghrifani et al., 2021; Said & Maryono, 2018). Mapingure et al. (2019) attest that travel motivation reveals tourists' needs, which leads to tourists behaving in a particular manner to achieve desired fulfillment. This study is in line with a recent study conducted by Katsikari et al. (2020), which posits that travel motivation is an important concept that explains tourists' behavior, motivation for visiting a particular destination, and overall travel satisfaction. Consequently, travel motivation has been recognized by Kara and Mkwizu (2020) as a key factor in determining tourist behavior. Therefore, this study advocates that tourist behavior at a destination is intimately tied to the motivations of tourists visiting that destination.

Kara and Mkwizu (2020) view travel motivation as an inner urge that propels an individual to choose a particular destination in the hope of achieving the desired benefits and satisfaction. According to Amuquandoh (2017), travel motivation influences tourists' preferences and can predict travel decisions by reflecting the type of experience, destination, or activity desired by tourists. Tourism researchers worldwide have used travel motivation to measure tourists' levels of satisfaction (Çelik & Dedeoğlu, 2019; Preko et al., 2019), establish an understanding of tourists' travel decisions and consumption patterns (Kara & Mkwizu, 2020; Yoo et al., 2018). Motivation remains elusive, although it has been extensively measured (Amuquandoh, 2017; Woyo & Slabbert, 2020). According to research, insufficient focus has been placed on determining the key "push and pull" factors, especially in SA where PAs are contributing to the SDGs, more especially to Goal 1 ("reducing poverty in all its forms"), Goal 8 ("economic growth"), and Goal 10 ("reduce inequality"). Governments have made many requests for a more in-depth understanding of the factors that affect visitor happiness to provide useful information for management decision-making (Agyeman et al., 2019; Casteel & Bridier, 2021).

PA managers must be better equipped with information to meet the needs of tourists and effectively market their products and services (Agyeman et al., 2019). A deeper understanding of travel motivation assists in the development of successful marketing strategies and policies to boost tourism demand (Kara & Mkwizu, 2020; Unguren et al., 2021). However, the core themes behind travel motivation revolve around "push" and "pull" factors. These factors have been extensively employed in tourism research to assess tourists' travel motivations. However, there is limited literature regarding why tourists continue to visit PAs in SA. Therefore, this study identifies key push and pull motives for tourists to continue visiting PAs. The current study fills this theoretical gap by investigating key travel motivations and statistically validating the relationship between push and pull motivation factors. The following section delves into push travel motivation factors related to this chapter.

8.2.2 Push Travel Motivation Factors

Push travel motivation factors (such as the need for escape, relaxation, knowledge and novelty seeking, prestige, and family/friends' togetherness) are linked to internal forces or sociopsychological aspects that influence travel decisions among tourists (Cengizci et al., 2020; Nejat & Fallah Lajimi, 2022). According to Subadra (2019), push factors are "the mental preparation for human actions and linked to an individual level of optimal arousal." Thus, the main motivation for tourists is their desire to travel (Subadra, 2019). In other words, this pertains to specific needs that tourists seek to satisfy when traveling and consider when choosing their destination (Aquino et al., 2019; Dean & Suhartanto, 2019). Consequently, Oyunchimeg and Gantuya (2021a) perceived push factors as the intangible, intrinsic, and origin-related desires of tourists. In the context of PAs, researchers (see Carvache-Franco et al., 2019; Karasakal & Dursun, 2018; Ma et al., 2018; Pereira et al., 2022; Wangari, 2017) have confirmed a variety of push travel motivation factors, such as the desire for leisure, adventure, relaxation, health, social and family interaction, escaping ordinary daily routine, sightseeing, and exploration. According to Karasakal and Dursun (2018), push factors drive tourists to travel to escape their hectic routines and to gain fresh experiences. The literature review shows that tourists are motivated by various push travel motivator factors. The following section explains the pull travel motivation factors related to the attractiveness of a destination as perceived by tourists.

8.2.3 Pull Travel Motivation Factors

Pull travel motivation factors are the external desires of tourists or the various elements of a destination, such as culture and natural attractions, that stimulate tourists' interest (Carvache-Franco et al., 2019). Pull travel motivation factors are the allure of a destination that draws tourists to visit it (Karasakal & Dursun, 2018), implying that they originate within the destination. Many researchers (for example, Aquino et al., 2019; Oyunchimeg & Gantuya, 2021b; Wambani et al., 2020) have outlined pull factors as tangible elements of a location, such as lifestyle, landscape, events, culture, heritage, or climate, that provide tourists with an external motive to travel. Hence, pull motivation factors are destination-specific characteristics that influence tourists' choice of one destination over another. Pull factors are critical for enhancing tourists' perceptions of the destination's worth and benefits (Aquino et al., 2019).

Previous studies (such as those of Aquino et al., 2019; Karasakal & Dursun, 2018; Oyunchimeg & Gantuya, 2021b; Wambani et al., 2020) have identified primary, secondary, and tertiary pull travel motivation factors. The primary pull factors can be classified as the history of the local community, scenery, climate, wildlife, historic sites, and local culture (Karasakal & Dursun, 2018). The secondary pull travel motivation factors focus on tourist services, which include accommodation, catering, personal attention and services, entertainment value, tourist accessibility, political stability of the destination, and trends and popularity of tourism destinations (Karasakal & Dursun, 2018; Wambani et al., 2020). Tertiary pull travel motivation factors include how the destination is marketed and the cost of traveling to that location (Amuquandoh, 2017). Nonetheless, Oyunchimeg and Gantuya (2021b) argue that pull travel motivation factors for PAs are likely to vary depending on the country, region, or location of the PA.

In this chapter, resources as pull factors include staff service delivery, infrastructure, information communication, and physical characteristics of the national park. According to Ezeuduji and Mgabhi (2023), HiP's high biodiversity, scenic beauty, history, culture, and surrounding communities have created many tourism opportunities for the park. Hence, HiP was used in this research as a PA. Eshun and Tichaawa (2020) attest that tourism in PAs is a mutually beneficial economic development strategy that benefits both the host communities and conservation. The next section presents the theories underpinning this study.

8.3 Theoretical Underpinning

The study reported in this chapter is based on two (2) theories: Maslow's hierarchy of needs theory (1943) and Dann's theory of push and pull motivations (1977). These frameworks provide useful insights for establishing an understanding of the relatizonships between push and pull motivational factors. Maslow's hierarchy of needs theory (1943) has been frequently applied in the field of tourism to explain tourists' travel behaviors (Yousaf et al., 2018). This study perceives Maslow's hierarchy of needs theory (1943) as human conduct classified into fulfilling various distinct levels of needs, with the satisfaction of one need leading to the quest to satisfy another at a higher level, resulting in a hierarchy.

Another generally acknowledged theoretical framework in tourism research is Dann's (1977) push and pull theory of motivation (see Mapingure et al., 2019; Said & Maryono, 2018). According to Dann (1977), various factors motivate visitors and/or tourists to visit destinations. These factors can be classified as pushes or pulls. This study, on the one hand, characterized push factors as internal drives that push individuals to travel. On the other hand, the study characterized pull factors as external factors that draw people to specific places or destinations (Michael et al., 2017; Said & Maryono, 2018). According to Said and Maryono (2018), these are related to elements such as longing for leisure, recreation, adventure, and escape. In this chapter, these two theories are employed as lenses to examine tourists' need for travel and travel motivations, and they are linked to the recommendations made to HiP management. In the field of tourism research, these theories are still relevant. Current research has conducted comparable studies using these as theoretical frameworks (Krishnapillai & Kwok, 2020; Maghrifani et al., 2021; Mapingure et al., 2019; Said & Maryono, 2018).

The researchers outlined aspects of the research design, data collection methods, and analyses in the following section.

8.4 Research Design

The study targeted tourists (both international and domestic) visiting HiP for a variety of reasons. For example, these reasons usually relate to push and pull travel motivation factors and hence are linked as contributing factors to enhance the studied constructs' variables. The positivist research paradigm for measurement was adopted as the study developed statistical relationships between push and pull travel motivation factors. This study used a probability sampling method, known as simple random sampling, in which the researchers randomly selected a subset of respondents from the research population, as the number of visitors to HiP could be estimated (Veal, 2011). A quantitative research approach was employed, as the nature of the study reported in this chapter required quantified information to arrive at conclusions, with the study research questions leaning toward a positivist research paradigm. Using a simple random sampling technique, 600 questionnaires were distributed between October 2022 and January 2023 to tourists who visited HiP, targeting different regions within the park. However, the researchers used 435 completed questionnaires that were usable for data analysis. To validate the sample size used in the study through statistical measurement, a significance level of 0.05 and −0.223 was used as the minimal absolute significant path coefficient, and 0.80 was employed as the power level. Considering these two methods (the inverse square root and gamma-exponential methods), the sample size generated for this study is statistically appropriate and substantial for 435 tourists who visited HiP.

Questionnaire variables included tourists' profile (categorical variables); push travel motivation factors—"relaxation, social interaction, family togetherness, and prestige"; pull travel motivation factors—"resources, recreational facilities, natural resources, history, culture, and learning experiences." On a 5-point Likert scale, all motivation factors were set as ordinal variables: 1 = strongly agree and 5 = strongly disagree. Motivation factor variables were derived from the literature reviewed.

For data analysis, partial least squares structural modeling (PLS-SEM) was conducted in two stages. The first stage was achieved using IBM's SPSS version 28 software (IBM Corp, 2021), while the second stage tested the measurement (which is construct validity and reliability) and structural model using composite-based structural equation modeling (SEM) involving a multivariate analytical technique implemented on WarpPLS software version 7.0. SEM represents a theory-driven data analytical approach for the evaluation of previously specified hypotheses regarding causal relations among measured and/or latent variables. Ultimately, the SEM process enables the evaluation of the fit between (typically) correlational data derived from experimental or non-experimental research (Hair et al., 2020).

Missing values were analyzed to address concerns stemming from incomplete data. Descriptive data analysis was performed to obtain the frequency of respondents' profiles (in percentages), mean scores (M), and standard deviations (SD) of the responses. The degree of internal consistency of the variables used to explain the different study dimensions of (the) push and (b) pull factors was assessed using Cronbach's alpha reliability test. Cronbach's alpha coefficients of 0.5–0.7 are acceptable in social science research (George & Mallery, 2003). The reliability test results in this research suggest that the variables employed to measure the motivation subdimensions are internally consistent. Pearson correlation coefficient tests (r) were used to assess the strength and direction of the relationship between the two variables. This chapter does not intend to dwell on statistical reasons, or the presentation of results packed with statistical indicators. However, essential statistical explanations or indicators were provided. Confirmatory factor analysis was conducted before the model measurement. The following section presents and discusses the key results of this study in relation to the research questions and their practical implications.

8.5 Presentation and Discussion of Findings

The research presented in this chapter explores the key push and pull factors influencing tourists visiting PAs. Respondents' profile variables are presented in the next subsection.

8.5.1 Respondents' Profile

The results show that more females (51.3%) than males (48.7%) responded to the questionnaire, and 55.4% of the respondents were mature or relatively old (more than 50 years of age). The result further reflects that there were more foreign nationals (about 52%) than South African respondents (about 48%). However, the percentages of tourists who were permanent and non-permanent residents in SA were almost equal. The study results further show that 69.2% of respondents surveyed were employed tourists, while 30.8% were unemployed, indicating that most tourists surveyed were working or had a job, assuming they had disposable income and thus could afford to travel.

Moreover, more than half of the respondents (approximately 63%) classified themselves as highly satisfied with their visit to HiP as a PA, and approximately 27% of the respondents classified themselves as moderately satisfied. In this study, 3.4% of respondents were neutral (neither satisfied nor dissatisfied) regarding their level of satisfaction with their visit. The results further suggest that 5.3% of respondents were moderately dissatisfied, and 1.8% of respondents were highly dissatisfied with their visit to the park. A recent study by Ezeuduji and Dlomo (2020) noted that tourists who are pleased with a destination are more prone to return and are more likely to become devoted to and recommend a destination to others, such as friends and family members. The following subsection responds to the first question.

Q1. What are the key tourists' "push factors" to visit PAs, such as HiP?

8.5.2 The Key Tourists' "Push" Factors to Visit PA in South Africa

The respondents were more positive than negative with statements on the four (4) push sub-dimensions—"relaxation, social interaction, family togetherness, and prestige motivations" measuring push travel motivation factors. The total mean score for each push sub-dimension was calculated to identify the key push factor(s) influencing tourists to visit the HiP. Relaxation, with a total mean score of 1.98, was identified as one of the key push factors influencing tourists to visit HiP, followed by family togetherness (2.41). The study reported in this chapter advocates that tourists visit HiPs primarily to relax and enjoy nature's peace and tranquility with their families. Thus, family travel is thought to be a particularly effective way to strengthen family ties and overall well-being. In support of these results, Michael et al. (2017) advocated that tourists who are pushed to travel and spend time with their families use vacations to establish and strengthen family values, make memories, and reconnect with their family members. HiP is a great destination for those seeking a peaceful retreat in a beautiful natural setting and for those who want to strengthen their family bonds. The following subsection responds to the second question.

Q2. What are the key tourists' "pull factors" to visit PAs, such as such as HiP?

8.5.3 The Key Tourists' 'Pull' Factors to Visit PA in South Africa

The respondents were more positive than negative with statements on the five (5) push sub-dimensions—"resources, learning experience, recreational facilities, natural resources, history and culture" measuring pull travel motivation factors. The results reveal that based on the calculated total mean scores for pull factors, "natural resources" (1.56), "destination resources" (1.81), "learning experience" (1.85), "history and culture" (1.95) were identified as key pull factors that influence tourists to visit HiP. The results indicate that tourists also visit HiP for educational purposes (learning experience), in addition to relaxation and spending time with family (push factors). The park is committed to conservation efforts and offers tourists the opportunity to learn about local flora and fauna (natural resources) and the challenges faced in conserving and preserving them. Tourists also learn about a destination's history and culture. Recent studies support these observations (for example, Karasakal & Dursun, 2018; Oyunchimeg & Gantuya, 2021b; Said & Maryono, 2018; Wambani et al., 2020).

The study reported in this chapter advocates that both intrinsic (push) and extrinsic (pull) factors influence tourists' decisions and motivation to travel; an analysis based on just one of these factors can only capture half of the reality (Faerber et al., 2021; Masina et al., 2021; Newland & Aicher, 2018; Whyte, 2017). Thus, when push and pull travel motivation factors are accounted for, the totality of tourist motivations is addressed. Understanding the "push and pull" factors that attract tourists to PAs at any destination is crucial. Drawing from Maslow's hierarchy of needs theory and Dann's theory of push and pull motivations, the findings of this study statistically validate that the identified key push (relaxation = 1.98; family togetherness = 2.41) and key pull (natural resources = 1.52; destination resources = 1.31; learning experience = 1.85; history and culture = 1.95) factors could assist in addressing the diversity of tourists' needs, and

Table 8.1 No significant correlation results for push and pull factors

Variables	Statistical indicators	Results
"Prestige motivations" (PM) and "relaxation" (RE)	$r = 0.04$, $p = 0.403$	No significant correlation
"Family togetherness" (FT) and "social interaction" (SI)	$r = 0.08$, $p = 0.94$	No significant correlation
"Natural resources" (NR) and SI	$r = 0.027$, $p = 0578$	No significant correlation
NR and PM	$r = 0.024$, $p = 0.619$	No significant correlation
"History and culture" (HC) and RE	$r = -0.048$, $p = 0.314$	No significant correlation
HC and FT	$r = -0.023$, $p = 0.626$	No significant correlation
HC and "resources" (RES)	$r = 0.016$, $p = 0.734$	No significant correlation

Source: Authors' compilation

the relationships between these needs and tourists' profiles are equally important for relationship marketing. This, in turn, assists destination management in identifying and satisfying tourists' needs by promoting and planning tourism offerings (products and services). Furthermore, a correlation test was used to determine the degree and direction of the relationship between the push and pull factors. Therefore, the following subsection responds to the third question mentioned earlier.

Q3. How is the relationship between "the push sub-factors" (relaxation, social interaction, family togetherness, and prestige motivations) and "pull sub-factors" (resources, learning experience, recreational facilities, natural resources, history, and culture) for tourists who visited the park, and is there a significant correlation between these two sub-dimensions?

8.5.4 Relationships Between Tourists' Push and Pull Factors to Visit PAs

A correlation coefficient analysis was performed to determine the degree and direction of the relationship between two or more variables. A significant linear relationship was found among most variables, although some showed a perfect negative correlation while others showed a perfect positive correlation. Variables that showed no significant correlation are shown in Table 8.1.

Most study variables (push and pull) exhibited significant linear relationships. Except for SI and RE, which showed a significant negative correlation (relaxation does not increase but decreases with social interaction), the other correlations were significant and positive (one variable enhanced the other). The variables are listed in Table 8.2.

Certain individual push and pull factors did not have a significant linear relationship. This may not hold true when composite constructs (push and pull factors) are tested for a linear relationship. The following subsection, therefore, responds to the fourth question:

Q4. How is the relationship between the composite push and pull factors for tourists who visited the park, and is there a significant correlation between these two study constructs?

The results presented a positive and significant linear relationship between the composite push and overall pull factors ($r = 0.501^{**}$, $p = 0.001$), and the pull factor increased with an increase in the push factor. The mean and standard deviation statistics also indicated a strong relationship between the two study constructs. There was a positive relationship between push (M = 42.03, SD = 7.782) and pull (M = 36.95, SD = 9.784) factors as indicated by the distribution of the points. These results are in line with other studies on travel motivation (Demirović et al., 2019; Kim & Lee, 2020; Mutanga et al., 2017; Wambani et al., 2020), which found a strong significant correlation between pull and push travel motivation factors in their studies. Notably, this correlation does not imply causation. There is no guarantee that one variable causes another because the two variables are correlated.

Table 8.2 Significant correlation results for push and pull factors

Variables	Statistical results	Variables	Statistical results
Significant negative correlation			
SI and RE	$r = -0.110^{*}, p = 0.022$	"Recreational facilities" (RF) and RE	$r = 0.323^{**}, p = <0.001$
Significant positive correlations			
FT and RE	$r = 0.438^{**}, p = <0.001$ (positive)	RF and SI	$r = 0.191^{**}, p = <0.001$
PM and SI	$r = 0.484^{**}, p = <0.001$	RF and FT	$r = 0.330^{**}, p = <0.001$
PM and FT	$r = 0.112^{*}, p = 0.019$	RF and PM	$r = 0.235^{**}, p = <0.001$
RES and RE	$r = 0.296^{**}, p = <0.001$	RF and RES	$r = 0.440^{**}, p = <0.001$
RES and SI	$r = 0.107^{*}, p = 0.026$	RF and LE	$r = 0.524^{**}, p = <0.001$
RES and FT	$r = 0.281^{**}, p = <0.001$	NR and RE	$r = 0.444^{**}, p = <0.001$
RES and PM	$r = 0.122^{*}, p = 0.011$	NR and FT	$r = 0.377^{**}, p = <0.001$
"Learning experience" (LE) and RE	$r = 0.348^{**}, p = <0.001$	NR and RES	$r = 0.333^{**}, p = <0.001$
LE and SI	$r = 0.301^{**}, p = <0.001$	NR and LE	$r = 0.455^{**}, p = <0.001$
LE and FT	$r = 0.152^{**}, p = 0.001$	NR and RF	$r = 0.521^{**}, p = <0.001$
LE and PM	$r = 0.222^{**}, p = <0.001$	HC and SI	$r = 0.414^{**}, p = <0.001$
LE and RES	$r = 0.270^{**}, p = <0.001$	HC and PM	$r = 0.196^{**}, p = <0.001$
HC and NR	$r = 0.142^{**}, p = 0.003$	HC and RF	$r = 0.281^{**}, p = <0.001$
HC and LE	$r = 0.448^{**}, p = <0.001$		

Source: Authors' compilation
$^{***}p < 0.001$, $^{**}p < 0.01$, and $^{*}p < 0.05$

Researchers (such as Amuquandoh, 2017; Inkson & Minnaert, 2018; Whyte, 2017) confirm that both push and pull factors exist on a continuum, with push factors having intangible attributes of the destination on one end and pull factors having tangible attributes on the other. In essence, tourists leave and travel to destinations because they are pushed and pulled by these two distinct forces (Dann, 1977).

As reported in this chapter, this study advocated that it is important for any PA that the management understands the motivations of their target markets (tourists/visitors) and then designs products and service offerings in harmony with these motivations. Furthermore, a positive relationship between push and pull factors is essential for HiP's marketing strategy as a tourism destination. This will be useful in addressing the diversity of tourists' needs and the relationships between them, as this is critical for relationship marketing.

8.6 Recommendations

Based on these findings, the following suggestions are made to the management and decision-makers of HiP. These recommendations help to demonstrate alternative approaches to accomplishing the goals of sustainable nature-based tourism, sustainable marketing, and promotion of visitation to PAs. The recommendations are organized into two themes or categories (marketing and promotion of HiP visits and strengthening visitor expectations).

8.6.1 Marketing and Promotion of Visitation to HiP

HiP marketing and promotion should be based on a well-informed strategy and a planned approach for sustainable visitation. This can be accomplished by considering the following:

- To reinforce visitor satisfaction behavior, carefully crafted communications emphasizing the identified key push (relaxation and family togetherness) and pull factors (natural resources, destination resources, learning experience, history, and culture) should be directed to the target market. Such promotional and communication initiatives can be offered on-site, online (including social media), and through advertisement/promotion campaigns using appropriate marketing platforms, with easily accessible information.
- The push and pull motivation factors should be evoked within the HiP's visual environment through promotional materials, travel agencies, positioning, and designed promotional programs to target these markets and create more innovative and needs-addressing packages for tourists (based on the identified push and pull factors) to stimulate tourists' eagerness to travel. HiP should present itself as a location for leisure and relaxation, social interaction, and family bonding by providing natural learning experiences, as well as history and culture packages in its amenities located in its rich natural setting and surroundings (park qualities). Consequently, facilities within and immediately outside the park should be designed to meet these needs. Rather than focusing on the overall mass market, park destination marketing initiatives should focus on specific markets (target marketing).

8.6.2 Enhancing Visitor Expectations

It is critical to meet tourist expectations because this leads to visitor satisfaction. This can be obtained by ensuring that

- Park operations tailor tourists' on-site service expectations (based on key push and pull factors) to meet or exceed their actual experiences.
- Park-service excellence practices must consider the quality consistency of information passed on to potential visitors about the park experience before their visit, the actual experience received by tourists when visiting the park, and post-sales service after the visitors have left the park.

The study reported in this chapter posits that tourists will continue to visit HiP to satisfy their internal and emotional elements (push factors) by engaging in the recognized park attributes and operations (pull factors).

8.7 Theoretical and Managerial Implications

This study contributes to the enhancement of the service rendered to tourists visiting PAs, using the marketing principles of identifying, understanding, and addressing the multiplicity of tourists' needs and the relationships between these needs, as this is critical for relationship marketing. This could assist in creating more innovative and needs-addressing packages for nature-based tourism (Wambani et al., 2020). This highlights the need for public-private partnerships, as well as relationships between specific parks and neighboring communities, in developing a strong network of stakeholders and developing more innovative and needs-addressing tourism packages (Eshun & Tichaawa, 2020). The study reported in this chapter suggests that the emphasis in PAs should not be merely transactional—solely on selling nature—as more can be done by park management to contribute to global agreements and frameworks such as the 2030 Agenda for Sustainable Development. According to the study findings, understanding consumers' wants and their characteristics is critical in satisfying visitor expectations, which must be realistic (Ma et al., 2018). Hence, the findings of this study should be of interest to scholars interested in understanding what constitutes tourist satisfaction within and between PAs, as this study contributes to the existing body of literature on tourist satisfaction in the South African context.

This research regarding novelty and contribution to theory thus conducted a more conclusive study and statistically modeled the push and pull travel motivation factors influencing tourists to visit PAs in KwaZulu-Natal, SA. It established not only how these factors influence each other regarding tourists' travel motivation to visit PAs in KwaZulu-Natal SA but also how they interact with each other to impact visit satisfaction. This study has the potential to fill a knowledge gap regarding the formulation of a conceptual model, validated by statistics, of the factors influencing tourists to visit PAs. Therefore, this study also provides an understanding of what constitutes tourists' satisfaction within and between PAs, as well as the literature on tourists' satisfaction in general.

8.8 Conclusion

This chapter highlights a recent study undertaken in the South African context that evaluated travel motivation (push and pull) for tourists visiting PAs. The findings suggest that "relaxation" and "family togetherness" are significant tourist push factors, whereas the key tourist pull factors highlighted include "natural resources, destination resources, learning experience, history and culture." Based on the study findings, it has been suggested that park management develop, promote, and sustain packages that address the identified key push and pull motivation factors. Although there is a positive linear relationship between the composite push and pull factors, the sub-dimensions of these two factors may not correlate. Therefore, emphasis must be placed on more holistic thinking regarding the relationship between motivational factors and park management, as opposed to the relationship between individual motivational factors.

8.9 Limitations and Future Research

The respondents identified in this study are tourists; further studies can also be conducted to determine which motivation factors or attributes are possessed by tourism-related employees (including park managers and other staff members) as internal customers in relation to visitor satisfaction (external customers). The significant influence and relationships between these factors and staff profiles can be beneficial for employee wellness and retention. Furthermore, other potential areas for future research could be the exploration of other areas (e.g., tourists' behavior, nostalgia, and motivation-based emotional arousal) in conjunction with motivation to develop a better understanding of the decision-making process of tourists regarding nature-based tourism in the South African context.

References

Abrahams, D. (2019). Transformation of the tourism sector in South Africa: A possible growth stimulant? *Geojournal of Tourism and Geosites, 26*(3), 821–830. https://doi.org/10.30892/gtg.26312-400

Agyeman, Y. B., Aboagye, O. K., & Ashie, E. (2019). Visitor satisfaction at Kakum National Park in Ghana. *Tourism Recreation Research, 44*(2), 178–189. https://doi.org/10.1080/02508281.2019.1566048

Amuquandoh, F. E. (2017). Tourists' motivations for visiting Kakum National Park, Ghana. *Ghana Journal of Geography, 9*(1), 152–168.

Aquino, R. S., Schänzel, H. A., & Hyde, K. F. (2019). Analysing push and pull motives for volcano tourism at Mount Pinatubo, Philippines. *Geoheritage, 11*(1), 177–191. https://doi.org/10.1007/s12371-017-0254-z

Carvache-Franco, M., Víquez-Paniagua, A. G., Carvache-Franco, O., Perez-Orozco, A., & Carvache-Franco, W. (2019). Motivations, intentions to return and to recommend protected areas: A study in Costa Rica. *Geo Journal of Tourism and Geosites, 27*(4), 1173–1183. https://doi.org/10.30892/gtg.27405-424

Carvalho, N., Rodrigues, H., & Brochado, A. (2023). Double or nothing: Push and pull factors of casinos in Europe. *GeoJournal, 88*(2), 2233–2246. https://doi.org/10.1007/s10708-022-10749-7

Casteel, A., & Bridier, N. L. (2021). Describing populations and samples in doctoral student research. *International Journal of Doctoral Studies, 16*(1), 339–362.

Çelik, S., & Dedeoğlu, B. B. (2019). Psychological factors affecting the behavioural intention of the tourist visiting South-eastern Anatolia. *Journal of Hospitality and Tourism Insights, 2*(4), 425–450. https://doi.org/10.1108/JHTI-01-2019-0005

Cengizci, A. D., Başer, G., & Karasakal, S. (2020). Exploring push and pull motivations of Russian tourists to Turkey. *Tourism Review International, 24*(2-3), 127–141. https://doi.org/10.3727/154427220X15912253254419

Dann, G. M. S. (1977). Anomie, ego-enhancement and tourism. *Annals of Tourism Research, 4*(4), 184–194. https://doi.org/10.1016/0160-7383(77)90037-8

Dean, D., & Suhartanto, D. (2019). The formation of visitor behavioural intention to creative tourism: The role of push-pull motivation. *Asia Pacific Journal of Tourism Research, 24*(5), 393–403. https://doi.org/10.1080/10941665.2019.1572631

Demirović, D., Berjan, S., Milentijević, N., El Bilali, H., & Syromiatnikova, Y. A. (2019). Exploration of tourist motivation and preferred activities in rural areas. *Journal of the Geographical Institute, 69*(1), 29–37.

Dube, K., & Nhamo, G. (2020). Tourism business operators' perceptions, knowledge and attitudes towards climate change at Victoria Falls. *The Journal for Transdisciplinary Research in Southern Africa, 16*(1), 1–10.

Eshun, G., & Tichaawa, T. M. (2020). Towards sustainable ecotourism development in Ghana: Contributions of the local communities. *Tourism: An International Interdisciplinary Journal, 68*(3), 261–277. https://doi.org/10.37741/t.68.3.2

Ezeuduji, I. O., & Dlomo, N. C. (2020). Push and pull factors for domestic leisure travel in Mtubatuba Local Municipality, South Africa. *EuroEconomica, 39*(2), 121–134.

Ezeuduji, I. O., & Mgabhi, N. Z. (2023). Potential for rural tourism development around Hluhluwe iMfolozi Park, South Africa: A local community perspective. *African Journal of Development Studies, 13*(1), 89. https://doi.org/10.31920/2634-3649/2023/v13n1a5

Faerber, L. S., Hofmann, J., Ahrholdt, D., & Schnittka, O. (2021). When are visitors actually satisfied at visitor attractions? What we know from more than 30 years of research. *Tourism Management, 84*, 104–284. https://doi.org/10.1016/j.tourman.2021.104284

George, D., & Mallery, P. (2003). *SPSS for Windows step by step: A simple guide and reference. 11.0 update*. Allyn & Bacon.

Hair Jr, J. F., Howard, M. C., & Nitzl, C. (2020). Assessing measurement model quality in PLS-SEM using confirmatory composite analysis. Journal of Business Research, 109, 101–110. https://doi.org/10.1016/j.jbusres.2020.11.069

IBM Corp. (2021). *IBM SPSS statistics for windows, version 28.0*. IBM Corp.

Inkson, C., & Minnaert, L. (2018). *Tourism management: An introduction*. Sage.

Kara, N. S., & Mkwizu, K. H. (2020). Demographic factors and travel motivation among leisure tourists in Tanzania. *International Hospitality Review, 34*(1), 81–103. https://doi.org/10.1108/IHR-01-2020-0002

Karasakal, S., & Dursun, A. (2018). Destination push and pull factors: A literature review. *Journal of Recreation and Tourism Research, 5*(1), 23–37.

Katsikari, C., Hatzithomas, L., Fotiadis, T., & Folinas, D. (2020). Push and pull travel motivation: Segmentation of the Greek market for social media marketing in tourism. *Sustainability, 12*(11), 1–18. https://doi.org/10.3390/su12114770

Kim, C., & Lee, S. (2020). Understanding the cultural differences in tourist motivation between Anglo-American and Japanese tourists. *Journal of Travel & Tourism Marketing, 9*(1-2), 153–170. https://doi.org/10.4324/9780203823989

Krishnapillai, G., & Kwok, S. Y. (2020). Uncovering the Muslim leisure tourists' motivation to travel domestically–do gender and generation matter? *Tourism and Hospitality Management, 26*(1), 213–231. https://doi.org/10.20867/thm.26.1.12

Le Corre, N., Saint-Pierre, A., Hughes, M., Peuziat, I., Cosquer, A., Michot, T., & Bernard, N. (2021). Outdoor recreation in French coastal and marine protected areas. Exploring recreation experience preference as a way for building conservation support. *Journal of Outdoor Recreation and Tourism, 33*, 1–11. https://doi.org/10.1016/j.jort.2020.100332

Leung, Y. F., Spenceley, A., Hvenegaard, G., Buckley, R., & Groves, C. (2018). *Tourism and visitor management in protected areas: Guidelines for sustainability*. IUCN.

Ma, A. T., Chow, A. S., Cheung, L. T., Lee, K. M., & Liu, S. (2018). Impacts of tourists' socio-demographic characteristics on the travel motivation and satisfaction: The case of protected areas in South China. *Sustainability, 10*(10), 1–21. https://doi.org/10.3390/su10103388

Maghrifani, D., Liu, F., & Sneddon, J. (2021). Understanding potential and repeat visitors' travel intentions: The roles of travel motivations, destination image, and visitor image congruity. *Journal of Travel Research, 61*(5), 1121–1137. https://doi.org/10.1177/00472875211018508

Mapingure, C., du Plessis, E., & Saayman, M. (2019). Travel motivations of domestic tourists: The case of Zimbabwe. *African Journal of Hospitality, Tourism and Leisure, 8*(2), 1–15.

Masina, N. O., Boshoff, L., & Sifolo, P. P. S. (2021). Wildlife tourism motivation: A case for Manyeleti Nature Reserve, Mpumalanga. *African Journal of Hospitality, Tourism and Leisure, 10*(4), 1414–1429. https://doi.org/10.46222/ajhtl.19770720-170

Michael, N., Wien, C., & Reisinger, Y. (2017). Push and pull escape travel motivations of Emirati nationals to Australia. *International Journal of Culture, Tourism and Hospitality Research, 11*(3), 274–296. https://doi.org/10.1108/IJCTHR-04-2016-0039

Mutanga, C. N., Vengesayi, S., Chikuta, O., Muboko, N., & Gandiwa, E. (2017). Travel motivation and tourist satisfaction with wildlife tourism experiences in Gonarezhou and Matusadona National Parks, Zimbabwe. *Journal of Outdoor Recreation and Tourism, 20*, 1–18. https://doi.org/10.1016/j.jort.2017.08.001

Nejat, S., & Fallah Lajimi, H. (2022). Identifying and prioritising factors affecting the revisit intention of the destination brand in Iran with the best-worst method. *Journal of International Marketing Modelling, 3*(1), 62–84. https://doi.org/10.22080/JIMM.2022.23408.1029

Newland, B. L., & Aicher, T. J. (2018). Exploring sport participants' event and destination choices. *Journal of Sport & Tourism, 22*(2), 131–149. https://doi.org/10.1080/14775085.2018.1436464

Nordbø, I., Mykletun, R. J., Segovia, J., & Segovia-Pérez, M. (2023). Push and pull motivations of international voluntary workers on organic farms (the WWOOFers). *Current Issues in Tourism, 26*(1), 137–152. https://doi.org/10.1080/13683500.2021.2004095

Oyunchimeg, L., & Gantuya, N. (2021a). Mongolian domestic tourists' motivation and revisit intention: Mediating effect of perceived benefit and perceived value. *Central European Journal of Geography and Sustainable Development, 3*(1), 32–48. https://doi.org/10.47246/CEJGSD.2021.3.1.3

Oyunchimeg, L., & Gantuya, N. (2021b). Understanding of travel motivations of domestic tourists. *Journal of Tourism and Services, 22*(12), 1–23. https://doi.org/10.29036/jots.v12i22.253

Pereira, V., Gupta, J. J., & Hussain, S. (2022). Impact of travel motivation on tourist's attitude toward destination: Evidence of the mediating effect of the destination image. *Journal of Hospitality and Tourism Research, 46*(5), 946–971. https://doi.org/10.1177/1096348019887

Preko, A., Doe, F., & Dadzie, S. A. (2019). The future of youth tourism in Ghana: Motives, satisfaction and behaviour intentions. *Journal of Tourism Futures, 5*(1), 5–21. https://doi.org/10.1108/JTF-12-2016-0059

Ramukumba, T. (2019). Community views on the role of tourism in local development: A South African study. *Geojournal of Tourism and Geosites, 25*(2), 638–647. https://doi.org/10.30892/gtg.25228-386

Said, J., & Maryono, M. (2018). Motivation and perception of tourists as push and pull factors to visit national parks. *E3S Web of Conferences, 31*, 1–5. https://doi.org/10.1051/e3sconf/20183108022

Subadra, I. N. (2019). Investigating push and pull factors of tourists visiting Bali as a world tourism destination. *International Journal of Multidisciplinary Educational Research, 8*(8), 253–269.

Unguren, E., Tekin, Ö., & Bayirli, M. (2021). Exploring the effect of push and pull motivation factors on destination satisfaction: An empirical evidence from amenity migration perspectives. *European Journal of Tourism Research, 28*, 1–22. https://doi.org/10.54055/ejtr.v28i.1946

United Nations. (2022). The sustainable development goals report 2022. Retrieved September 29, 2023, from https://unstats.un.org/sdgs/report/2022/

Veal, A. J. (2011). *Research methods for leisure and tourism: A practical guide* (4th ed.). Financial Times Prentice Hall.

Wambani, P. P., Ogunjinmi, A. A., & Oladeji, S. O. (2020). Socio-demographic determinants of travel motivation and behaviour of visitors in nature-based destinations in northern Nigeria. *Journal of Applied Sciences and Environmental Management, 24*(12), 2121–2129. https://doi.org/10.4314/jasem.v24i12.18

Wangari, C. (2017). Understanding the travel motivation among youth travellers in Kenya: the 'push' and 'pull' paradigm. *African Journal of Hospitality, Tourism and Leisure, 6*(1), 1–16.

Whyte, L. J. (2017). Understanding the relationship between push and pull motivational factors in cruise tourism: A canonical correlation analysis. *International Journal of Tourism Research, 19*(5), 557–568. https://doi.org/10.1002/jtr.2129

World Bank Group. (2017). *Tourism for development: 20 reasons sustainable counts for development*. The World Bank.

Woyo, E., & Slabbert, E. (2020). Unpacking the motivations, satisfaction and loyalty of tourists travelling to a distressed destination. *Anatolia, 31*(4), 536–548. https://doi.org/10.1080/13032917.2020.1794919

Ying, K. S., & Krishnapillai, G. (2018). Young domestic tourists—Are they motivated to travel? *Advanced Science Letters, 24*(5), 3444–3449. https://doi.org/10.1166/asl.2018.11405

Yoo, C. K., Yoon, D., & Park, E. (2018). Tourist motivation: An integral approach to destination choices. *Tourism Review, 73*(2), 169–185. https://doi.org/10.1108/TR-04-2017-0085

Yousaf, A., Amin, I., & Santos, C. J. A. (2018). Tourist's motivations to travel: A theoretical perspective on the existing literature. *Tourism and Hospitality Management, 24*(1), 197–211. https://doi.org/10.20867/thm.24.1.8

9 Visitor Experiences in Kruger National Park Camps: A Sentiment Analysis of Camp Reviews Post-COVID-19

Kaitano Dube

Abstract

National parks play a vital role in conserving biodiversity and promoting tourism; however, the rise in tourism raises concerns about its environmental impact. This study was conducted during the COVID-19 pandemic, significantly altering tourists' preferences and behaviours. To assess visitor sentiments towards these camps, the study used sentiment analysis of 164 online reviews from Trip Advisor. The findings were positive, with high ratings for camp locations, particularly those near riverbanks that offered breathtaking wildlife views. However, challenges with infrastructure maintenance, service quality, and cleanliness were noted, particularly in older camps such as Pretoriuskop. This study recommends enhancing maintenance schedules, improving service delivery, and upgrading infrastructure to meet modern standards. Additionally, it suggests integrating renewable energy solutions and improving Internet connectivity to enhance the overall tourist experience. Addressing the challenges various camps face is crucial in ensuring that Kruger National Park maintains a competitive edge over competition from private players within and outside the park. To achieve this, park management must commit seriously to capital investment, given the existing tourism market. The study recommends improved customer engagement on social media platforms to address potential service failure in the post-COVID-19 era. These insights are crucial for park management in planning and improving visitor experiences at Kruger National Park.

Keywords

Service excellence · Nature-based tourism · Service failure · Green tourism · Park hospitality · Sustainable tourism · COVID-19

9.1 Background and Introduction

National parks play a central role in nature-based tourism hubs worldwide, serving as key tourist destinations that offer a mixture of biodiversity conservation, recreational opportunities, and critical environmental education (Silva et al., 2023). Given the unique cultural and natural resources of protected areas, several tourists are attracted to national parks annually (Dube et al., 2022; Dube & Nhamo, 2020). Africa has several national parks of global significance as they assist in

K. Dube (✉)
Department of Tourism and Integrated Communication, Vaal University of Technology, Vanderbijlpark, South Africa

Emirates Aviation University, Dubai Academic City, Dubai, United Arab Emirates

K. Dube et al. (eds.), *Tourism in Rural Protected Areas*, Sustainable Development Goals Series,
https://doi.org/10.1007/978-3-031-87333-1_9

addressing several sustainable development goals (SDGs) (Mabibibi et al., 2021). Among other issues, national parks support the protection of biodiversity conservation and contribute to SDG 15—life on land. National Parks equally contribute to several other SDGs linked to tourism as critical tourism hubs where they employ several thousands of people and contribute to the poverty alleviation of host communities.

However, a surge in tourism in national parks is often a cause of concern for conservation practitioners and other stakeholders as it can potentially trigger environmental degradation (Mearns et al., 2023). This has been evident in various places worldwide, such as Indonesia's mountainous national parks, where unregulated tourism has resulted in pollution, habitat deterioration, and other negative impacts (Pamungkas & Jones, 2021). In summary, national parks are crucial for nature-based tourism, providing essential services for ecological processes and habitats and offering tourism and educational experiences (Burgess, 2014). However, the sustainability of these tourism hubs is contingent on effective management and regulation to mitigate environmental impacts and ensure biodiversity conservation (Pamungkas & Jones, 2021). Addressing these challenges is essential to the long-term viability of national parks as nature-based tourism destinations.

Most of these parks are located in sub-Saharan Africa (SSA). These national parks act as critical conservation hubs but are equally important; these areas act as major tourist destinations that attract tens of thousands of visitors annually. Major tourism hubs in the form of national parks include Serengeti National Park in Tanzania, Maasai Mara National Game Reserve in Kenya, Kruger National Park in South Africa, Chobe National Park in Botswana, Hwange National Park in Zimbabwe, and Etosha National Park in Namibia. As protected areas, the main national parks are home to natural heritage sites in the form of flora and fauna (biodiversity) (Buchling & Maroun, 2021; Chikodzi et al., 2022). They are held in high esteem by the public, particularly ecotourists. These national parks are also home to cultural heritage sites with a very long history, dating back hundreds and hundreds of years. Some national parks boast beautiful scenery views and the so-called Big 5 animals, which separate the major national parks in Sub-Saharan Africa from other national parks.

Kruger National Park in South Africa is a critical tourism, heritage, and biodiversity asset and one of the most visited national parks in the country and region (Dube & Nhamo, 2020; Liang et al., 2020; Matshusa et al., 2021; Wessels & Douglas, 2022). Its dominance and influence have far-reaching impacts and contribute to the surrounding communities. There is consensus that the success of any tourism enterprise depends on the treatment of guests or tourists (Dewi & Abidin, 2021; Engeset et al., 2016; Singh et al., 2024). In this regard, the issue of customer service excellence is crucial to the success and development of tourism destinations. Tourism destinations and tourism managers often invest a substantial amount of time in analysing and understanding tourists' preferences (Kovačić et al., 2022), which are key drivers of success. In doing so, understanding customer satisfaction is key for tourism marketers (Kim et al., 2020; Orden-Mejía et al., 2022).

Regardless of this knowledge, customer service studies in Africa and South Africa are far in between. This robs marketers of key information they can use to drive tourism development, including customer-centric tourism products and services. Where such studies exist, they are dated (Engelbrecht et al., 2014; Leberman & Holland, 2005; Scholtz et al., 2015), given that tourist preferences are, at best, dynamic, and forever evolving, as such studies need constant updating to ensure the timely delivery of new and emerging tourist trends. Post-COVID-19, such studies are a must, as the pandemic significantly altered tourists' preferences, attitudes, and how they consume tourism products and services (Asyahri et al., 2023; Borges-Tiago et al., 2021; Bramwell, 1998; León et al., 2023).

While a few studies have been conducted by various scholars on tourist satisfaction in Kruger

National Park (Chihwai, 2019; Kruger et al., 2015; Nkwanyana et al., 2018; Slabbert, 2018), this is the first study conducted post the COVID-19 pandemic to gain insights as to how the pandemic could have altered tourists satisfaction levels. Unlike other studies, this study used online tourist reviews. By default, the study will provide some of the key challenges picked with regard to tourist satisfaction in Kruger National Park before the pandemic. This study aimed to analyse visitor reviews to understand the overall sentiment towards different camps in Kruger National Park. The key research to this study is as follows: What is the overall sentiment expressed by visitors towards different camps in Kruger National Park? This study is critical in informing some of the key emerging visitor preferences in national parks, particularly Kruger National Parks. This will assist in informing tourism practitioners of appropriate tools for planning and camping accommodation design and planning. The study's findings will provide valuable insights into the factors that contribute to positive and negative visitor experiences and can help inform strategic decisions to enhance the overall quality of the park's tourism offerings.

Online reviews are increasingly used to assess tourist sections across the tourism sector (Gretzel & Yoo, 2008; Gursoy, 2019; Nilashi et al., 2018; Sotiriadis & Van Zyl, 2013). The review provides holistic yet subjective views of the tourists. Platforms such as Google Reviews can provide critical insights into tourists' views on various aspects of tourism product offerings and, as such, are at the centre of tourism service excellence delivery. These platforms can be integrated into customer feedback, providing an inexpensive way to gain customer feedback, and managers need to consider views from such platforms as tourists use them in deciding on whether to visit various destinations. Many tourism establishments have put integrators in place to link customer feedback to platforms such as Trip Advisor. User-generated data is considered valuable in tourism decision-making (Alsayat, 2023; Carpio et al., 2024; Ji et al., 2023).

9.2 Methodology

The study adopted a case study research approach, which examined data from four randomly selected camps from Kruger National Parks: Skukuza Camp, Pretoriuskop, Olifants, and Letaba Camp. Olifants and Letaba camps are located in the northern sections of the camp, whereas the other two are located in the southern section of the camp. The Skukuza Camp is the main camp, which also doubles as the main administrative section of the park. Pretoriuskop and Skukuza are two of the oldest and first established camps with the longest history in the park, which was established in 1928 (SANParks, 1990). The Olifants Camp was built in 1960, and the Letaba Camp was built much earlier in 1947. These selected camps are ideal for the study and are deemed to be representative of the sentiment of tourists visiting the KNP.

The study relied on the analysis of customer reviews on Trip Advisor for the period between 2020 and 2024. In this regard, 164 reviews from the four camps were analysed. The first step was to copy all review comments from each campsite's platform. Soon after copying the reviews into a Word document, a cleaning process ensued to ensure that the copied documents only comprised comments for review. In addition, language editing was performed to ensure that the document was read as fluently as possible. Subsequently, the four camp reviews were loaded into ATLAS.ti for processing purposes. The questions outlined in Table 1 were used to generate codes for this study (Fig. 9.1).

Consequently, 68 codes were generated and 149 quotations were equally developed. The codes were grouped into nine thematic areas, covering amenities and service impact, cleanliness and maintenance impact, customer service drivers, hospitality, hut conditions, impact on the environment, restaurant quality, wildlife, and nature experience on tourist satisfaction. The sentiment analysis function was prompted, starting with an analysis of the positive sentiment analysis and later followed by the negative sentiment analysis to analyse all the data that was fed

Fig. 9.1 Questions used for code generation in data analysis using ATLAS.ti

into the ATLAS.ti system. The top applied codes are shown in Fig. 9.2.

9.3 Results and Discussions

The study found general and positive sentiment about the camps in Kruger National Park experience, with tourists ranking the camp facilities between 3.5 and 4.0 on a scale of 1–5, with 1 being the lowest quality and 5 being the highest quality (Fig. 9.2). The three camps are ranked 4, while the Pretoriuskop rest camp is the weakest among all the camps. The highest-ranked parameter by visitors is the location for Skukuza, Olifants, and Letaba, which scored 4.5, demonstrating that these camps are best located at face value. The camps are located on riverbanks, offering guests a picturesque view of wildlife. Pretoriuskop was ranked lower than the other camps at 4.0 for this parameter. Generally, in terms of location, the four camps are ideal and are located in a good environment inside Kruger National Park (Fig. 9.3). Located near Numbi Gate, Pretoriuskop, however, does not have a river view, unlike the other three camps.

Regarding cleanliness, Skukuza Rest Camp took the lead at 3.5, and the other camps were tied at 3.0. Therefore, cleanliness at all camps was above average. The camps' services are above average, and tourist rates are 3.5 for Skukuza and Letaba camps, a rating of 4 for Olifants and 3.5 for Pretoriuskop Camp.

Camps provide various services, facilities, and areas of interest which are critical for tourist enjoyment during their stay, ranging from restaurants, swimming pools, braai, and cooking facilities to allow for self-catering, among others.

9.4 Drivers of Positive Sentiments at Kruger National Park Camps

A sentiment analysis found that a number of tourists were happy with the camps, with 1419 quotations having a positive sentiment in terms of their experience with their park experience. Tourists highlighted various aspects of their experiences in Kruger National Park camps, with wildlife nature experience at the top of their list as a critical driver of their park's positive experiences.

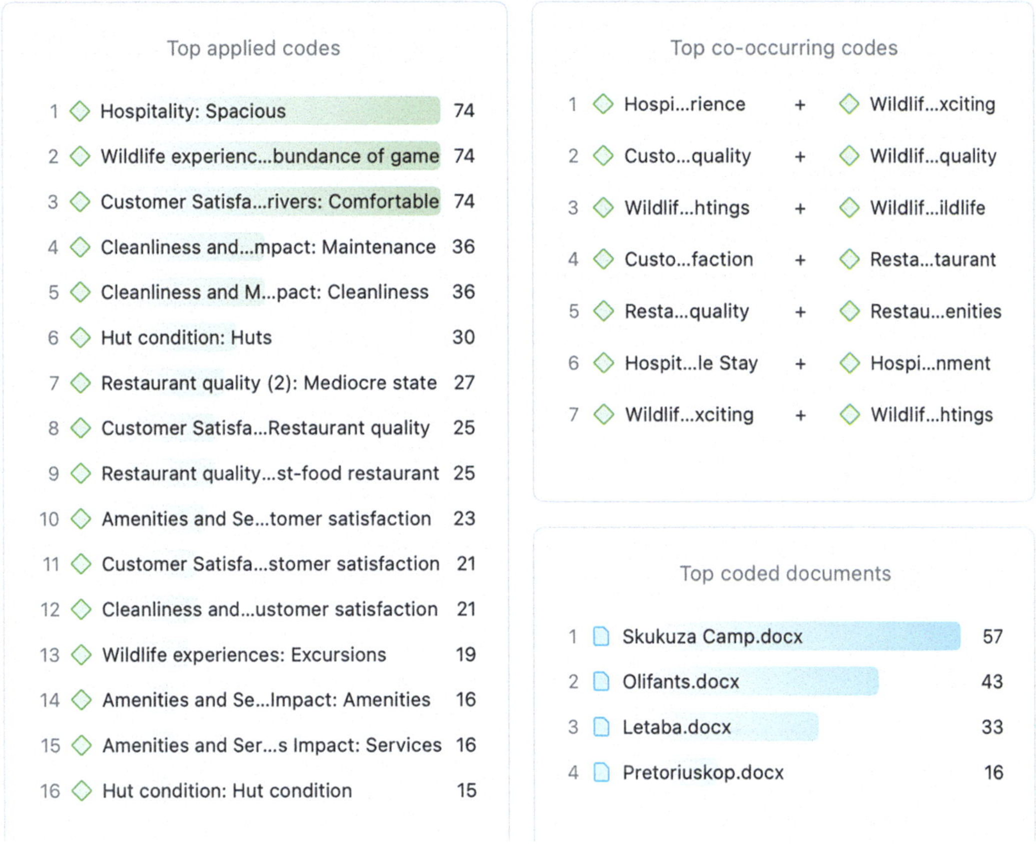

Fig. 9.2 Top applied codes and top occurring codes

They also applauded the service experience as part of the vital component that excited them during their visits to various camps in Kruger National Park. An extract from the sentiment analysis shows that tourists (Fig. 9.4), for example, to the Letaba camp, are particularly thrilled by the hospitality experience at the camp, which is complemented by game viewing served by hospitable staff.

The positive experience is not limited to dining services but to other facets of the camp, which they say is located in a serene area with good views, which allows guests to enjoy a wide variety of game and tree species in the area. Tourists mentioned a wide variety of animal spotting, which they enjoyed during their stay at the camp. Complimenting wildlife enjoyment requires a clean accommodation facility and informative guided tours.

Guests of the Olifants River Camp were equally optimistic about their experience, noting similar issues with the other camp at Letaba Camp, such as the beautiful river gorge scenery, good food experience at the restaurant, a serene and quiet environment which allows tourists to enjoy a beautiful landscape, and some which form part of the Big 5 (Fig. 9.5). They equally complement the friendliness of the staff and clean facilities in a non-crowded environment as key drivers of their positive experiences at their camps.

Tourists were equally excited about the tourist experience, and Pretoriuskop was accredited for its great reception, clean bathrooms, and other associated facilities (Fig. 9.6). In Skukuza, tourists indicated that various accommodation facilities influenced part of the positive experience. The camp offers several

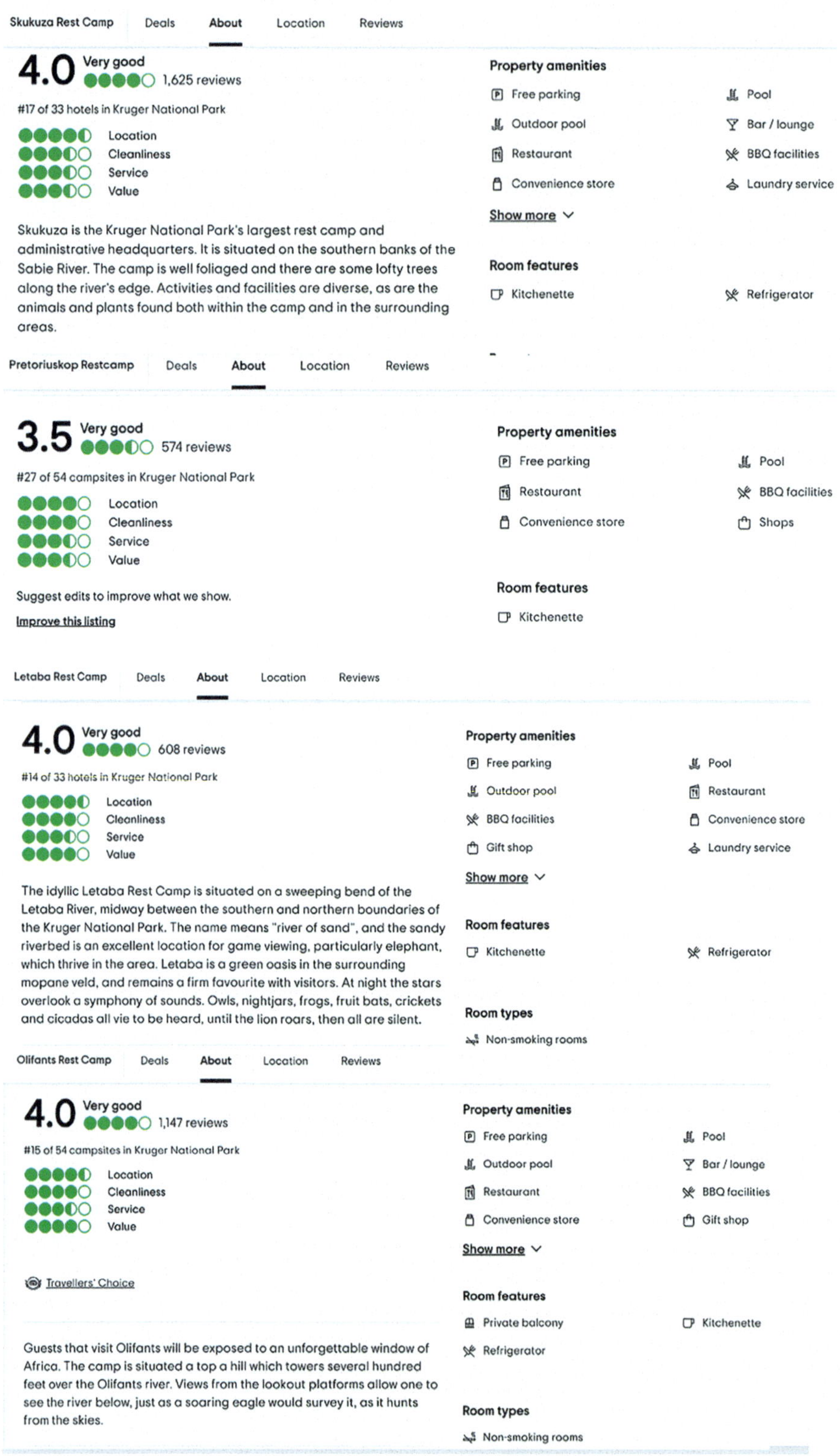

Fig. 9.3 Trip advisor reviews of the four selected camps in Kruger National Park

1:13 ¶ 68 – 69 in *Letaba.docx*

6. [illegible]

We only swung by here for lunch on a day trip from Olifants but were absolutely blown away by the view of the river from the restaurant terrace: we sat with a glass of wine, having lunch, and watching seven hippos emerge from the river to graze. The camp had a really lovely feel to it and I'd like to stay there next time we visit....

1:14 ¶ 78 – 80 in *Letaba.docx*

Pleasant, helpful staff
The new shop is open and looks great.
Busy rest camp stayed in the basic chalet with a shared kitchen.
beds comfortable, fridge inside unit, kettle and toaster in room, a short extension cord is useful as there is a plug by the fridge and one at the basin....

1:18 ¶ 91 – 93 in *Letaba.docx*

8. Hi guys,
It's a quiet spot. With a spectacular view over the Letaba River.
Last year I saw a pack of dogs in the riverbed.
This year, there were just kudus, elephants, waterbucks, impalas, storks, and thogs.
the restaurant serves tasty food. And the atmosphere is so relaxing....

1:22 ¶ 115 – 117 in *Letaba.docx*

10. [illegible] wrote a review Jul 2023
Canberra, Australia520 contributions243 helpful votes
Some great wildlife viewing
This is one of our regular Kruger Rest Camp choices in a great location on the Letaba River. It was sad to see on our last visit that many of the amazing fig trees which made the place so go...

1:26 ¶ 134 in *Letaba.docx*

There was a brilliant view of the Lataba River, and the accommodation was clean and very comfortable. The staff at reception, the restaurant and the shop were friendly and helpful. We went on a Sunset guided tour which was very informative. We also took a guide for a night drive and saw a Lion chomping on a Buffalo
Read more...

Fig. 9.4 Extract showing some of the key drivers of positive experience at Letaba Camp

accommodation facilities to suit the needs of various tourism markets, including bungalows and high-end market accommodation establishments, such as the Skukuza Safari Lodge, which opened weeks before the COVID-19 pandemic in early 2020.

Well-stocked shop facilities provide much-needed convenience for the public. The shops at the camp offer a variety of goods and other essential supplies, including gas. In addition, tourists were commended for good WIFI and network availability at the camp. The staff at Skukuza is also noted to be friendly, which adds to the good ambience and spacious accommodation.

9.5 Drivers of Negative Tourist Satisfaction in Kruger National Park Camps

Apart from the positive experience, tourists had some challenging issues that park management had to pay attention to. The study revealed 955 quotations with negative sentiments from the reviewer's comments. Concerning the experiences of tourists at the Letaba camp, the tourists noted disturbing incidences of non-working equipment, dirty toilet facilities, and rusting and dated furniture, which require attention. Given the age of some of the facilities, there might be a need to constantly check the facilities and

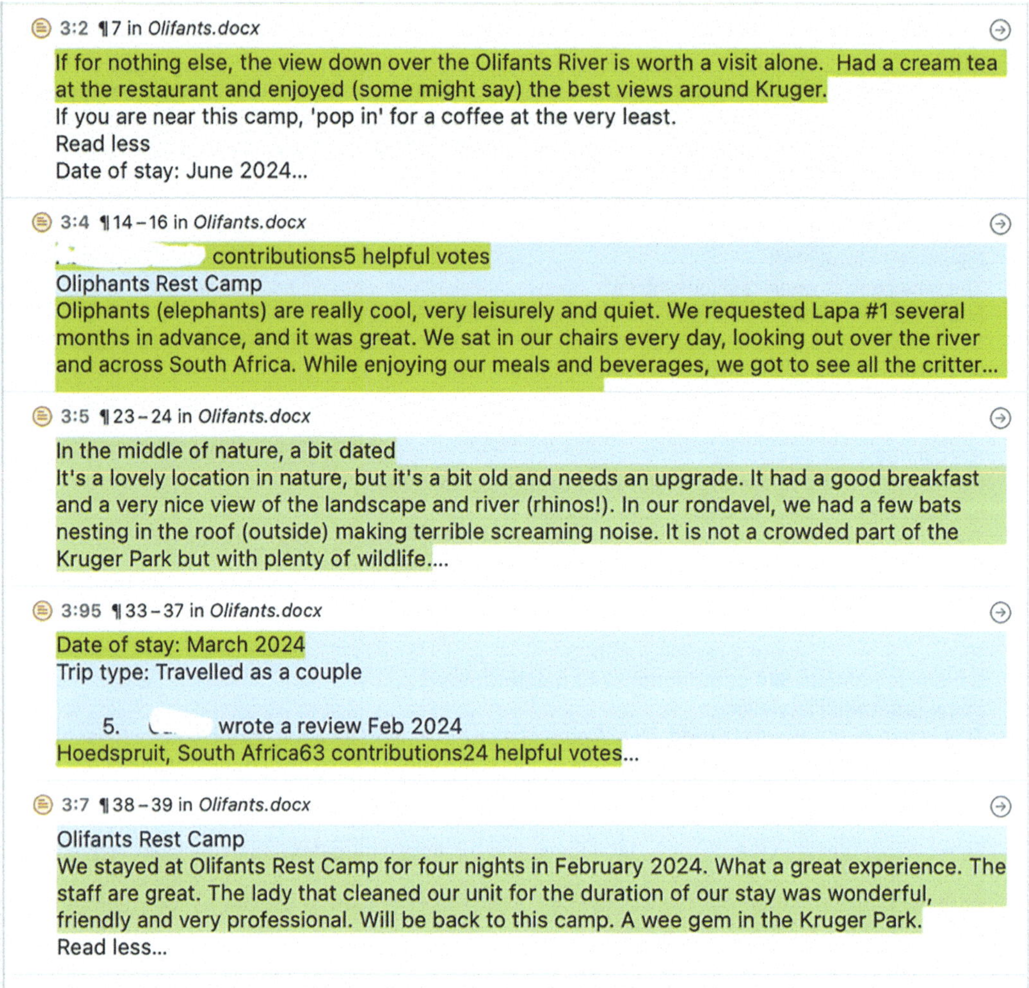
3:2 ¶7 in *Olifants.docx*

If for nothing else, the view down over the Olifants River is worth a visit alone. Had a cream tea at the restaurant and enjoyed (some might say) the best views around Kruger.
If you are near this camp, 'pop in' for a coffee at the very least.
Read less
Date of stay: June 2024...

3:4 ¶14–16 in *Olifants.docx*

contributions5 helpful votes
Oliphants Rest Camp
Oliphants (elephants) are really cool, very leisurely and quiet. We requested Lapa #1 several months in advance, and it was great. We sat in our chairs every day, looking out over the river and across South Africa. While enjoying our meals and beverages, we got to see all the critter...

3:5 ¶23–24 in *Olifants.docx*

In the middle of nature, a bit dated
It's a lovely location in nature, but it's a bit old and needs an upgrade. It had a good breakfast and a very nice view of the landscape and river (rhinos!). In our rondavel, we had a few bats nesting in the roof (outside) making terrible screaming noise. It is not a crowded part of the Kruger Park but with plenty of wildlife....

3:95 ¶33–37 in *Olifants.docx*

Date of stay: March 2024
Trip type: Travelled as a couple

5. wrote a review Feb 2024
Hoedspruit, South Africa63 contributions24 helpful votes...

3:7 ¶38–39 in *Olifants.docx*

Olifants Rest Camp
We stayed at Olifants Rest Camp for four nights in February 2024. What a great experience. The staff are great. The lady that cleaned our unit for the duration of our stay was wonderful, friendly and very professional. Will be back to this camp. A wee gem in the Kruger Park.
Read less...

Fig. 9.5 Key drivers of customer positive experience at Olifants Camp, Kruger National Park

upgrade them to ensure that they are up to date and in correct conditions and working order. In some instances, infrastructure maintenance is key to ensuring the optimal operation of equipment and in-room technology for the convenience of tourists (Fig. 9.7).

These challenges are not unique to the Letaba camp but extend to other camps. Pretoriuskop faces several challenges that require urgent attention by management, including infrastructure maintenance, which is the biggest issue. This could be attributed to the dated infrastructure, with maintenance falling behind in several respects. Furnishings, doors, and the ground require attention. Tourists also complained about poor meals served only by the restaurant that serves the camp. Tourists also complained of noise from staff accommodations, disturbing the beautiful environment's tranquillity.

Some challenges tourists cite point to laxity in supervision and camp management, with hospitality in mind. One of the tourists also complained about a lack of privacy, as someone photographed them while they were in the shower. This indicates a serious invasion of privacy and borders on sexual offences. There is also a need to relook the design of the camp huts, as some guests complained of being in cramped conditions (Fig. 9.8).

Tourists to Skukuza Rest Camp complained equally of the dated facility, lack of WIFI in

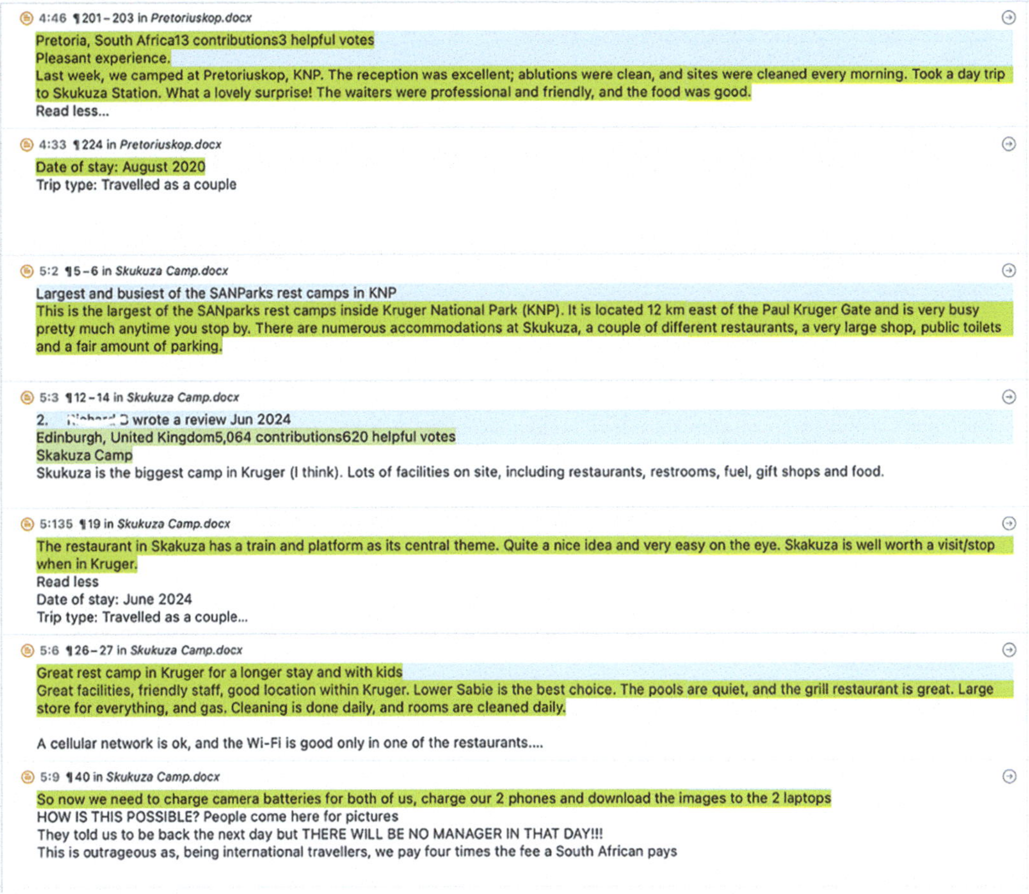

Fig. 9.6 Tourists' sentiment of Pretoriuskop and Skukuza Rest Camp

rooms, slow service at restaurants, and an urban feel-like environment which robs tourists of the bush feel they would want to get when out in the park, such as Kruger National Park. Tourists also complained about poor infrastructure maintenance, with some doors in a state of disrepair and unable to close properly. Tourists complained of not having electricity during load shedding owing to a lack of backup power. In 2023, South Africa faced serious challenges due to an energy shortage, which crippled the economy (Dube & Nhamo, 2024).

Some staff members also complained that they could not offer service excellence to tourists, with some accused of being unfriendly to tourists. Other tourists also complained of expensive services, while others called for an even more diverse food menu for some restaurants in the camp.

9.6 Discussions and Practical Implications

9.6.1 Critical Success Drivers for Tourism in Protected Areas

The study highlighted several critical success drivers in Kruger National Park's camps that ensure customer satisfaction. Amongst other things, the study demonstrated that in many respects, regardless of the challenges faced during the past decades by many national parks in Sub-Saharan Africa, ranging from climate change (Saarinen et al., 2022) to the harsh impacts of the

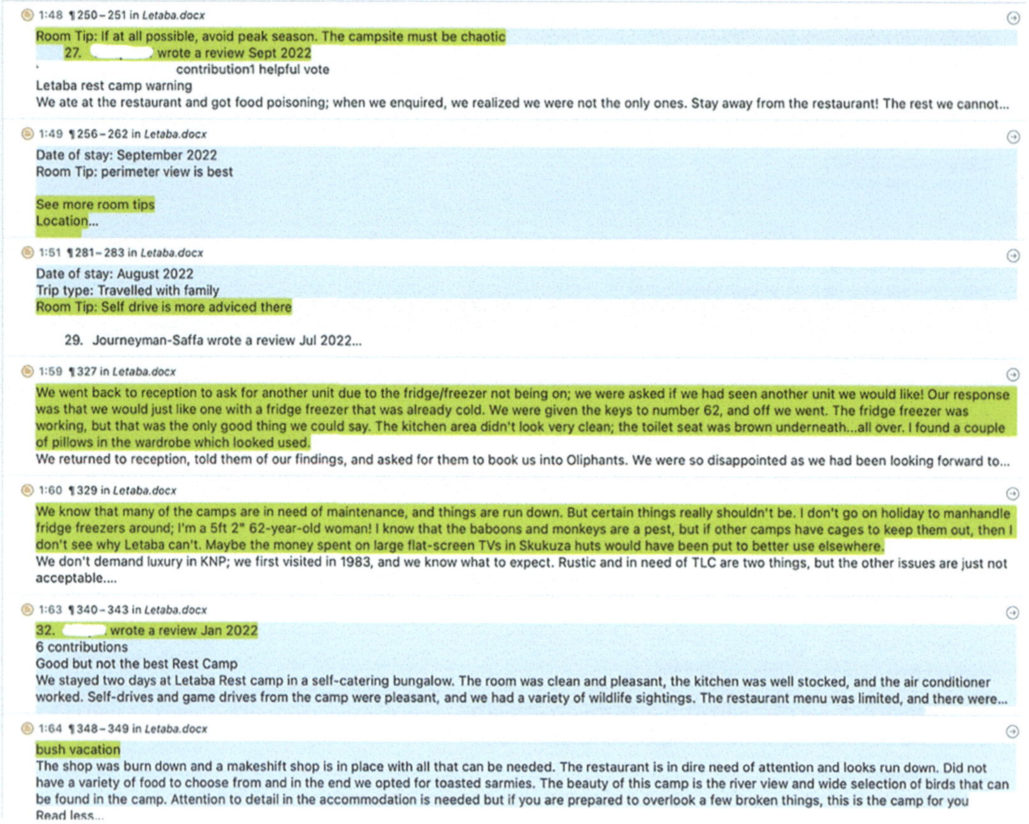
1:48 ¶250–251 in *Letaba.docx*
Room Tip: If at all possible, avoid peak season. The campsite must be chaotic
27. wrote a review Sept 2022
contribution1 helpful vote
Letaba rest camp warning
We ate at the restaurant and got food poisoning; when we enquired, we realized we were not the only ones. Stay away from the restaurant! The rest we cannot...

1:49 ¶256–262 in *Letaba.docx*
Date of stay: September 2022
Room Tip: perimeter view is best

See more room tips
Location...

1:51 ¶281–283 in *Letaba.docx*
Date of stay: August 2022
Trip type: Travelled with family
Room Tip: Self drive is more adviced there

29. Journeyman-Saffa wrote a review Jul 2022...

1:59 ¶327 in *Letaba.docx*
We went back to reception to ask for another unit due to the fridge/freezer not being on; we were asked if we had seen another unit we would like! Our response was that we would just like one with a fridge freezer that was already cold. We were given the keys to number 62, and off we went. The fridge freezer was working, but that was the only good thing we could say. The kitchen area didn't look very clean; the toilet seat was brown underneath...all over. I found a couple of pillows in the wardrobe which looked used.
We returned to reception, told them of our findings, and asked for them to book us into Oliphants. We were so disappointed as we had been looking forward to...

1:60 ¶329 in *Letaba.docx*
We know that many of the camps are in need of maintenance, and things are run down. But certain things really shouldn't be. I don't go on holiday to manhandle fridge freezers around; I'm a 5ft 2" 62-year-old woman! I know that the baboons and monkeys are a pest, but if other camps have cages to keep them out, then I don't see why Letaba can't. Maybe the money spent on large flat-screen TVs in Skukuza huts would have been put to better use elsewhere.
We don't demand luxury in KNP; we first visited in 1983, and we know what to expect. Rustic and in need of TLC are two things, but the other issues are just not acceptable....

1:63 ¶340–343 in *Letaba.docx*
32. wrote a review Jan 2022
6 contributions
Good but not the best Rest Camp
We stayed two days at Letaba Rest camp in a self-catering bungalow. The room was clean and pleasant, the kitchen was well stocked, and the air conditioner worked. Self-drives and game drives from the camp were pleasant, and we had a variety of wildlife sightings. The restaurant menu was limited, and there were...

1:64 ¶348–349 in *Letaba.docx*
bush vacation
The shop was burn down and a makeshift shop is in place with all that can be needed. The restaurant is in dire need of attention and looks run down. Did not have a variety of food to choose from and in the end we opted for toasted sarmies. The beauty of this camp is the river view and wide selection of birds that can be found in the camp. Attention to detail in the accommodation is needed but if you are prepared to overlook a few broken things, this is the camp for you
Read less...

Fig. 9.7 Service challenges faced by Letaba camp

COVID-19 pandemic (Mashapa & Dube, 2023), which started in 2020, the park is still an indomitable tourist destination that delivers an iconic wildlife experience. Tourists indicated their enjoyment of various animal species. Tourists demonstrated epic memories of their experience with flora and fauna, including the so-called Big 5 in different parts of Kruger National Park.

Apart from offering wildlife, this study found that Kruger National Park offers several add-on experiences for tourists, including accommodation and unforgettable cuisine. The study found an intricate link between good hospitality offerings in national parks and other service offerings in the park and how such interaction can enhance tourists' experience and build nostalgia, enhancing the destination image. Understanding this intricate link is crucial for destination management. Therefore, destination managers must have the skills and knowledge to manage these areas.

The study also noted that, apart from ensuring vibrant hospitality service offerings in the form of food and accommodation, tourists, besides being in the bush, still want to enjoy internet connectivity to connect. A couple of tourists intimated that WIFI connectivity and cellphone connectivity are critical for their enjoyment and, as such, ensuring this connectivity is central to tourism development in the future. Reliable Internet connections are essential for protected areas to allow tourists to connect with the world, and sharing their memories and experiences is a critical aspect of tourism product consumption in national parks. This view is supported by Walden-Schreiner et al. (2018) those who highlighted the centrality of the Internet in protected areas elsewhere in the world.

Tourists also highlighted the centrality of an innovative approach to the wildlife game viewing experience and the centrality of having a knowl-

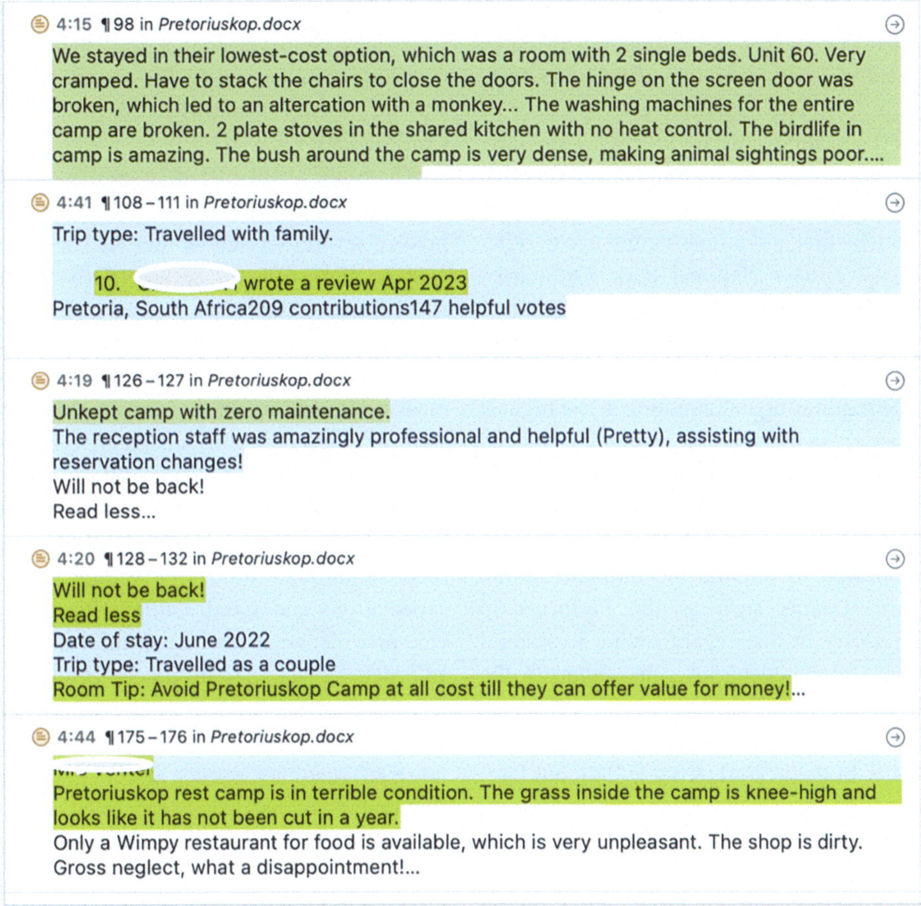
4:15 ¶ 98 in *Pretoriuskop.docx*

We stayed in their lowest-cost option, which was a room with 2 single beds. Unit 60. Very cramped. Have to stack the chairs to close the doors. The hinge on the screen door was broken, which led to an altercation with a monkey... The washing machines for the entire camp are broken. 2 plate stoves in the shared kitchen with no heat control. The birdlife in camp is amazing. The bush around the camp is very dense, making animal sightings poor....

4:41 ¶ 108 – 111 in *Pretoriuskop.docx*

Trip type: Travelled with family.

10. [illegible] wrote a review Apr 2023

Pretoria, South Africa209 contributions147 helpful votes

4:19 ¶ 126 – 127 in *Pretoriuskop.docx*

Unkept camp with zero maintenance.

The reception staff was amazingly professional and helpful (Pretty), assisting with reservation changes!

Will not be back!

Read less...

4:20 ¶ 128 – 132 in *Pretoriuskop.docx*

Will not be back!

Read less

Date of stay: June 2022

Trip type: Travelled as a couple

Room Tip: Avoid Pretoriuskop Camp at all cost till they can offer value for money!...

4:44 ¶ 175 – 176 in *Pretoriuskop.docx*

[illegible]

Pretoriuskop rest camp is in terrible condition. The grass inside the camp is knee-high and looks like it has not been cut in a year.

Only a Wimpy restaurant for food is available, which is very unpleasant. The shop is dirty. Gross neglect, what a disappointment!...

Fig. 9.8 Tourists' concerns regarding their stay at Pretoriuskop camp

edgeable guide. Tourists indicated their enjoyment of bush braai and commented on the role of interpretation by guides in enhancing their game-drive experience. Such reviews provide critical insights into what continues in a protected area to make the experience memorable.

9.6.2 Challenges and Root Causes of Service Failures in Some of the Camps in Kruger National Park

Kruger National Park is one of Africa's largest national parks in terms of operations. Given the age of the park, dating back more than a century, it has some dated infrastructure which requires attention to bring the infrastructure up to date. It is critical to note that when the park imagined accommodation as early as the early 1920s, it had considered and factored in the needs of tourists and clientele that were in existence then. Such recognition is critical for maintaining current and future tourist demands. This acknowledgement can assist tourism managers in rethinking accommodation design and technology to bring the same up-to-date.

Old infrastructure can be expensive for service and repair. The study noted that a generic complaint for all four camps revealed similar challenges of old infrastructure, which is now in a state of disrepair. Reading comments from tourists also points to a lack of a proper maintenance schedule, which can be evidence of the absence of a property management system to deal with the camps' challenges. A combination of dated infrastructure, laxity in terms of maintenance, and a lack of maintenance plans is a source of discomfort and annoyance for tourists. This study confirms earlier findings by Slabbert (2018) that there was a link between infrastructure age and tourist satisfaction and infrastructure age in other parts of the Kruger National Park. Ordinarily, when tourists are on vacation, they do not want to be annoyed by challenges that come along with assertions and infrastructure malfunctioning. This issue requires urgent attention at the highest level of park management. There might be a need to significantly upgrade or redo some old infrastructure to ensure much more energy-efficient, water-efficient buildings that incorporate green building design to provide the ultimate tourist experience. Camps such as the Pretoriuskop require priority in this regard owing to several complaints compared to any other camp in the park.

Maintenance challenges must be addressed holistically. In this regard, Kruger National Park must invest in a property management system that integrates all hotel operations with the capability to single out maintenance requirements. Various software packages can be deployed to deal with these complex tasks, which are tried and tested and already used by other hoteliers in the private sector and move away from the manual ways of doing things.

As part of operational design, enhancing security at Pretoriuskop might need to do away with public showers and build more ensuite sets to ensure the privacy and security of tourists, and the public facilities should only be left to campers. In particular, the Pretoriuskop camp has to enforce quiet hours and ensure that the staff is well-disciplined in setting up noise levels to ensure the comfort of tourists and the disturbance of animals.

There also seem to be inconsistencies in the cleaning of rooms. While most tourists praised the camps as clean areas, complaints to the minority might be evidence of cleaning standards by various staff members, and management and supervision should be proactive in addressing these disparities, which could be evidence of some lazy elements among staff members. The centrality of cleaning to the tourist experience was previously acknowledged in a study by Chihwai (2019).

There are also isolated cases of poor service offerings from restaurants in the park, which involve several service failures such as poor food quality, slow pace of service delivery of purchased meals, and isolated cases of food poisoning. These are issues which can tarnish the destination, and as such, there must be corrective measures to address them. Continuous training of staff and the availability of attentive supervisors can assist in alleviating some of these challenges. There must be improved communication between the customer and the restaurant with regard to food turnaround times for orders to manage expectations and avoid conflict. Restaurant service providers can share such information when an order is placed. To do this accurately, benchmarking and data analytics must be employed.

In addition, there is a need to foster a shared sense of customer service excellence, which is championed by parks and other third-party stakeholders to ensure uniformity and provide excellent service for tourists. Proper calculation of turnaround times and benchmarking are required to ensure service delivery.

With the ever-decreasing cost of renewables (Osman et al., 2023; Sayed et al., 2023; Weinand et al., 2023), Kruger National Park and SANParks have to enhance their bid towards a green energy transition to ensure that it is not only responsive to the needs of the current tourists who demand green tourism products (Gunduz Songur et al., 2022; Nhamo et al., 2023; Nowacki et al., 2023). This ensures that camps are energy-secure and carbon-neutral. The redesigning and rethinking of camps must ensure that the roofs are solar-ready, and some parking spaces can also be turned into solar spaces to deal with the space challenge in some respects. There is equally a need to address the growing need for carbon-neutral game-drive experience in line with the growing demand for sustainability (Kifworo et al., 2024).

9.7 Conclusion

The present study aimed to investigate the tourist experience at four selected camping facilities situated in the Northern and Southern regions of the Kruger National Park. The findings revealed that although tourists are generally content with the tourism products provided at the various camps, camping within the Kruger National Park presents several advantages. These include the exceptional experience of night game drives and prime viewing points for wildlife. For example, game viewing, cleanliness, staff attention to detail, quality of infrastructure, and access to services such as WIFI and game viewing experience are critical success drivers for tourists' experience in Kruger National Park post the COVID-19 pandemic. Additionally, the camps offer a unique opportunity to dine and wine at some of the park's restaurants, where experience with staff is critical to achieving customer satisfaction. Despite the camps' numerous infrastructure and facilities designed to enthuse tourists, several challenges commonly encountered within the camps can negatively impact the overall tourist experience in the park. These challenges encompass substandard service and accommodation offerings and inadequate catering. To mitigate these challenges, it is recommended that management supervision be improved and capital investments be made to address issues related to outdated and failing infrastructure. Given the centrality of platforms such as TripAdvisor, SANParks must develop social media spaces to deal with customer complaints and queries as a way of customer engagement and, equally importantly, as a platform to address service failure.

9.8 Study Limitations

The study is based on postings by travellers who sometimes do not always report accurately on events, and, as such, some information posted on such cannot be entirely relied on, although it provides a general view of tourists' experiences reflecting thrills and frustrations. Tourists' views cannot always be treated as gospel truth, as the account is always one-sided, and bias cannot be ruled out. The author distances himself from tourists' views in this regard.

References

Alsayat, A. (2023). Customer decision-making analysis based on big social data using machine learning: A case study of hotels in Mecca. *Neural Computing and Applications, 35*(6), 4701–4722. https://doi.org/10.1007/s00521-022-07992-x

Asyahri, Y., Wafa, F. E., Syaugi, S., & Muhaimin, M. (2023). Multiplier effect of religious tourism on income of the surrounding community in Banjar regency. *International Journal of Tourism and Hospitality in Asia Pacific, 6*(2), 2. https://doi.org/10.32535/ijthap.v6i2.2456

Borges-Tiago, T., Silva, S., Avelar, S., Couto, J. P., Mendes-Filho, L., & Tiago, F. (2021). Tourism and COVID-19: The show must go on. *Sustainability, 13*(22), 22. https://doi.org/10.3390/su132212471

Bramwell, B. (1998). User satisfaction and product development in urban tourism. *Tourism Management, 19*(1), 35–47. https://doi.org/10.1016/S0261-5177(97)00091-5

Buchling, M., & Maroun, W. (2021). Biodiversity reporting practices of the South African national parks. *Social Responsibility Journal, 19*(1), 138–165. https://doi.org/10.1108/SRJ-06-2020-0264

Carpio, A. J., Laguna, E., Pascual-Rico, R., Martínez-Jauregui, M., Guerrero-Casado, J., Vicente, J., Soriguer, R. C., & Acevedo, P. (2024). Prohibiting recreational hunting of wild ungulates in Spanish National Parks: Challenges and opportunities. *Science of the Total Environment, 926*, 171363. https://www.sciencedirect.com/science/article/pii/S0048969724015031

Chihwai, P. (2019). Developing SAFSERV: A comprehensive framework to measure tourists' satisfaction in Kruger National Park. *Business & Social Sciences Journal, 4*(2), 29–51.

Chikodzi, D., Nhamo, G., Dube, K., & Chapungu, L. (2022). Climate change risk assessment of heritage tourism sites within South African national parks. *International Journal of Geoheritage and Parks, 10*(3), 417–434. https://doi.org/10.1016/j.ijgeop.2022.08.007

Dewi, A. N., & Abidin, Z. (2021). Analysis of the relationship of service quality, motivation and destination image to destination loyalty: A case study of Wonorejo Mangrove ecotourism in Surabaya, East Java. *Journal of Aquaculture and Fish Health, 10*(1), 1. https://doi.org/10.20473/jafh.v10i1.19922

Dube, K., & Nhamo, G. (2020). Evidence and impact of climate change on South African national parks. Potential implications for tourism in the Kruger National Park. *Environmental Development, 33*, 100485. https://doi.org/10.1016/j.envdev.2019.100485

Dube, K., & Nhamo, G. (2024). Tourism resilience and challenges in Limpopo, South Africa: A post-COVID-19 analysis. *Development Southern Africa, 41*, 686–703. https://doi.org/10.1080/0376835X.2024.2334891

Dube, K., Nhamo, G., & Chikodzi, D. (2022). Climate change-induced droughts and tourism: Impacts and responses of Western Cape Province, South Africa. *Journal of Outdoor Recreation and Tourism, 39*, 100319. https://doi.org/10.1016/j.jort.2020.100319

Engelbrecht, W. H., Kruger, M., & Saayman, M. (2014). An analysis of critical success factors in managing the tourist experience at Kruger National Park. *Tourism Review International, 17*(4), 237–251. https://doi.org/10.3727/154427214X13910101597120

Engeset, M. G., Hull, J. S., & Velvin, J. (2016). Promoting service excellence for tourist destinations. *International Journal of Culture, Tourism and Hospitality Research, 10*(4), 440–454. https://doi.org/10.1108/IJCTHR-08-2015-0097

Gretzel, U., & Yoo, K. H. (2008). Use and impact of online travel reviews. In P. O'Connor, W. Höpken, & U. Gretzel (Eds.), *Information and communication technologies in tourism*. Springer. https://doi.org/10.1007/978-3-211-77280-5_4

Gunduz Songur, A., Turktarhan, G., & Cobanoglu, C. (2022). Progress on green technology research in hotels: A literature review. *Journal of Hospitality and Tourism Insights, 6*(5), 2052–2072. https://doi.org/10.1108/JHTI-10-2021-0280

Gursoy, D. (2019). A critical review of determinants of information search behavior and utilization of online reviews in decision making process (invited paper for 'luminaries' special issue of International Journal of Hospitality Management). *International Journal of Hospitality Management, 76*, 53–60.

Ji, F., Cao, Q., Li, H., Fujita, H., Liang, C., & Wu, J. (2023). An online reviews-driven large-scale group decision making approach for evaluating user satisfaction of sharing accommodation. *Expert Systems with Applications, 213*, 118875.

Julian, Bayliss Marije, Schaafsma Andrew, Balmford Neil D., Burgess Jonathan M.H., Green Seif S., Madoffe Sana, Okayasu Kelvin S.-H., Peh Philip J., Platts Douglas W., Yu (2014) The current and future value of nature-based tourism in the Eastern Arc Mountains of Tanzania Ecosystem Services 875–83 https://doi.org/10.1016/j.ecoser.2014.02.006

Kifworo, C., Ndhlovu, E., & Dube, K. (2024). Tourist perspectives on the use of electric cars in game lodges in Sub-Saharan Africa. In *Tourism and hospitality for sustainable development*. Springer. https://doi.org/10.1007/978-3-031-63073-6_3

Kim, G., Duffy, L. N., & Moore, D. (2020). Tourist attractiveness: Measuring residents' perception of tourists. *Journal of Sustainable Tourism, 28*(6), 898–916. https://doi.org/10.1080/09669582.2019.1708919

Kovačić, S., Jovanović, T., Vujičić, M. D., Morrison, A. M., & Kennell, J. (2022). What shapes activity preferences? The role of tourist personality, destination personality and destination image: Evidence from Serbia. *Sustainability, 14*(3), 3. https://doi.org/10.3390/su14031803

Kruger, M., Saayman, M., & Slabbert, E. (2015). Managing visitors' dining and retail experiences in South African national parks. *South African Journal of Business Management, 46*(2), 43–53. https://doi.org/10.10520/EJC172099

Leberman, S. I., & Holland, J. D. (2005). Visitor preferences in Kruger National Park, South Africa: The value of a mixed-method approach. *Journal of Park and Recreation Administration, 23*(2), 21–35.

León, C. J., González Hernández, M. M., & Lam-González, Y. (2023). COVID-19 effects on travel choices under climate risks. *Annals of Tourism Research, 103*, 103663. https://doi.org/10.1016/j.annals.2023.103663

Liang, Y., Kirilenko, A. P., Stepchenkova, S. O., & Ma, S. (2020). Using social media to discover unwanted behaviours displayed by visitors to nature parks: Comparisons of nationally and privately owned parks in the Greater Kruger National Park, South Africa. *Tourism Recreation Research, 45*(2), 271–276. https://doi.org/10.1080/02508281.2019.1681720

Mabibibi, M. A., Dube, K., & Thwala, K. (2021). Successes and challenges in sustainable development goals localisation for host communities around Kruger National Park. *Sustainability, 13*(10), 10. https://doi.org/10.3390/su13105341

Mashapa, M. M., & Dube, K. (2023). Tourism recovery strategies from COVID-19 within national parks in Western Cape, South Africa. In K. Dube, I. Mensah, & L. Chapungu (Eds.), *COVID-19, tourist destinations and prospects for recovery: volume two: An African perspective* (pp. 205–223). Springer International Publishing. https://doi.org/10.1007/978-3-031-24655-5_11

Matshusa, K., Leonard, L., & Thomas, P. (2021). Challenges of geotourism in South Africa: A case study of the Kruger National Park. *Resources, 10*(11), 11. https://doi.org/10.3390/resources10110108

Mearns, K., Dube, K., & Donkor, F. K. (2023). Locking in lessons from the COVID-19 pandemic for tourism destinations: A case study of wildlife destinations. In K. Dube, O. L. Kupika, & D. Chikodzi (Eds.), *COVID-19, tourist destinations and prospects for recovery: volume three: A South African and Zimbabwean perspective* (pp. 153–165). Springer International Publishing. https://doi.org/10.1007/978-3-031-28340-6_9

Nhamo, G., Dube, K., Chapungu, L., & Chikodzi, D. (2023). Quest for NetZero emissions in South African national parks: A tourism perspective. *Heliyon, 9*(6), e16410. https://doi.org/10.1016/j.heliyon.2023.e16410

Nilashi, M., Ibrahim, O., Yadegaridehkordi, E., Samad, S., Akbari, E., & Alizadeh, A. (2018). Travelers decision making using online review in social network sites: A case on TripAdvisor. *Journal of Computational Science, 28*, 168–179.

Nkwanyana, S., Bekebu, U., & Mhlongo, S. (2018). Evaluating tourists' expectations and experiences visiting the royal natal national park and game reserves in Southern drakensberg, KwaZulu-Natal. *African Journal of Hospitality, Tourism and Leisure, 7*(4), 1–11.

Nowacki, M., Kowalczyk-Anioł, J., & Chawla, Y. (2023). Gen Z's attitude towards green image destinations, green tourism and behavioural intention regarding green holiday destination choice: A study in Poland and India. *Sustainability, 15*(10), 10. https://doi.org/10.3390/su15107860

Orden-Mejía, M., Carvache-Franco, M., Huertas, A., Carvache-Franco, W., Landeta-Bejarano, N., & Carvache-Franco, O. (2022). Post-COVID-19 tourists' preferences, attitudes and travel expectations: A study in Guayaquil, Ecuador. *International Journal of Environmental Research and Public Health, 19*(8), 8. https://doi.org/10.3390/ijerph19084822

Osman, A. I., Chen, L., Yang, M., Msigwa, G., Farghali, M., Fawzy, S., Rooney, D. W., & Yap, P.-S. (2023). Cost, environmental impact, and resilience of renewable energy under a changing climate: A review. *Environmental Chemistry Letters, 21*(2), 741–764. https://doi.org/10.1007/s10311-022-01532-8

Pamungkas, W., & Jones, T. E. (2021). Indonesia's mountainous protected areas: National parks and nature-based tourism. In T. E. Jones, H. T. Bui, & M. Apollo (Eds.), *Nature-based tourism in Asia's mountainous protected areas: A trans-regional review of peaks and parks* (pp. 111–131). Springer International Publishing. https://doi.org/10.1007/978-3-030-76833-1_6

Saarinen, J., Fitchett, J., & Hoogendoorn, G. (2022). *Climate change and tourism in Southern Africa*. Routledge.

SANParks. (1990). *Tourism history – Kruger National Park*. SANParks. https://www.sanparks.org/parks/kruger/explore/tourism-history

Sayed, E. T., Olabi, A. G., Alami, A. H., Radwan, A., Mdallal, A., Rezk, A., & Abdelkareem, M. A. (2023). Renewable energy and energy storage systems. *Energies, 16*(3), 1415.

Scholtz, M., Kruger, M., & Saayman, M. (2015). Determinants of visitor length of stay at three coastal national parks in South Africa. *Journal of Ecotourism, 14*(1), 21–47. https://doi.org/10.1080/14724049.2015.1023730

Silva, S., Silva, L. F., & Vieira, A. (2023). Protected areas and nature-based tourism: A 30-year bibliometric review. *Sustainability, 15*(15), 15. https://doi.org/10.3390/su151511698

Singh, S. V., Ancheary, T., Mondal, A., Rajauria, S., Singh, S. V., Ancheary, T., Mondal, A., & Rajauria, S. (2024). Enhancing service excellence in the hospitality industry through effective service recovery from failures. In *Innovative technologies for increasing service productivity*. IGI Global. https://doi.org/10.4018/979-8-3693-2019-8.ch007

Slabbert, L. (2018). Co-creation in the Kruger National Park: Customer-driven improvement and satisfaction of the Lower Sabie safari tented accommodation. *Collaboration & Co-Creation Opportunities in Tourism, 12*, 146.

Sotiriadis, M. D., & Van Zyl, C. (2013). Electronic word-of-mouth and online reviews in tourism services: The use of twitter by tourists. *Electronic Commerce Research, 13*, 103–124.

Walden-Schreiner, C., Leung, Y.-F., & Tateosian, L. (2018). Digital footprints: Incorporating crowd-sourced geographic information for protected area management. *Applied Geography, 90*, 44–54. https://doi.org/10.1016/j.apgeog.2017.11.004

Weinand, J. M., Hoffmann, M., Göpfert, J., Terlouw, T., Schönau, J., Kuckertz, P., McKenna, R., Kotzur, L., Linßen, J., & Stolten, D. (2023). Global LCOEs of decentralized off-grid renewable energy systems. *Renewable and Sustainable Energy Reviews, 183*, 113478.

Wessels, J.-A., & Douglas, A. (2022). Exploring creative tourism potential in protected areas: The Kruger National Park Case. *Journal of Hospitality and Tourism Research, 46*(8), 1482–1499. https://doi.org/10.1177/1096348020983532

Part IV

Tourism in Rural and Protected Areas Conclusion and Recommendations

Redefining Tourism in Rural and Protected Areas: A Practical and Policy Perspective

10

Magdalena Petronella Swart, Kaitano Dube, and Ikechukwu O. Ezeuduji

Abstract

Rural tourism and ecological diversity in protected areas have gained significant research interest. It is reshaping the potential of sustainable community development despite persistent economic, natural and socio-cultural challenges. This chapter summarises some of the key messages, recommendations, practical implications, and limitations of this focus area. The lack of policies and regulations hampers the economic development of rural social entrepreneurs. Marginalised communities, such as women and youth, can reap substantial benefits through increased financial investment, capacity building, skills development, and education. Pro-poor tourism is highlighted as a vital approach to poverty alleviation, especially if local community farmers are integrated into the supply chain of wildlife and protected areas. Transparent management and community participation are the cornerstones of rural tourism product development, economic growth, and sustainable tourism. Marketing strategies informed by visitor management tools enhance visitor satisfaction, leading to positive word of mouth to promote rural destinations. Future research needs to focus on methodological improvements and practical policy implementations as proposed by the Southern African Framework for Rural Tourism and Protected Areas. Governments, social entrepreneurs, and community leaders have a critical role to play in the advancement of rural tourism development and the conservation of protected areas.

Keywords

Astrotourism · Protected areas · SDGs · Sustainable tourism · Microenvironment · Governance

10.1 Introduction

Southern Africa's vast terrain remains largely rural, with many areas covered by protected areas in the form of protected areas. In this regard, pro-

M. P. Swart (✉)
Department of Applied Management, University of South Africa, Pretoria, South Africa
e-mail: Swartmp@unisa.ac.za

K. Dube
Department of Tourism and Integrated Communication, Vaal University of Technology, Vanderbijlpark, South Africa

Emirates Aviation University, Dubai Academic City, Dubai, United Arab Emirates
e-mail: kaitanod@vut.ac.za

I. O. Ezeuduji
Department of Recreation and Tourism, University of Zululand, KwaDlangezwa Campus, Zululand, South Africa
e-mail: ezeudujii@unizulu.ac.za

K. Dube et al. (eds.), *Tourism in Rural Protected Areas*, Sustainable Development Goals Series,
https://doi.org/10.1007/978-3-031-87333-1_10

tected areas and rural tourism are central cogs for development in this world region, which has remained largely underdeveloped in many respects. Remote areas in this region have several iconic tourist destinations, not only in the form of national parks, which have been at the epicentre of wildlife tourism (Dube, Chikodzi, et al., 2023; Dube, Kupika, & Chikodzi, 2023), but also other destinations such as deserts, river destination areas, mountain destination sites and some of the pristine, breathtaking beaches, and wetlands which have helped in shaping tourism and tourism activities in some of the most resort areas. In Tanzania (Jiwaji, 2016), Namibia (Stone, 2019), and South Africa, remote areas such as Sutherland and Carnarvon towns in the Northern Cape offer unique tourism experiences and potential in the form of astrotourism (Ingle, 2010; Jacobs et al., 2020; Wassenaar & Coetzee, 2024). This area offers a new and unique tourism experience, given the vantage and remoteness of the area, which allows for various sky-viewing tourism experiences at the Square Kilometre Array (SKA) and the Large Telescope in Sutherland. This has opened up that area, which has been famous for decades for producing the Karoo lamb and slow economic growth. There is equally a huge potential to market some of these areas as hubs for education tourism with the potential to attract domestic and international tourists. This is particularly so given the huge investment in radio telescopes (Ingle, 2010; Jacobs et al., 2020; Wassenaar & Coetzee, 2024), which have made impressive discoveries and established this remote area as an astrophysics and science hub (Weltman et al., 2020).

Even with such a promise for socioeconomic development and potential, rural tourism remains challenged by several factors, amongst them infrastructural development, due to low investment and priority by the government. Access roads, poor internet connectivity, poor access to clean water infrastructure, disaster outbreaks (Dube, Kupika, & Chikodzi, 2023), climate change (Dube, 2024), and criminality (Ferreira, 1999; Minnaar & Herbig, 2018; Ntuli & Potgieter, 2001; Tan & Nguwi, 2023) are critical challenges facing rural and protected areas in remote Southern Africa.

Despite the challenges tourism faces in rural and protected areas over the past decades, the emphasis on rural tourism has grown slowly (Fig. 10.1). This is curious, particularly given the centrality of that kind of tourism and the various themes that can be covered under this topic. Wide knowledge gaps, therefore, still exist in this kind of tourism spectrum, which requires focus and attention. Any success and development of tourism enterprise in Southern Africa has to be squarely focused on the understanding of urban tourism and rural tourism. The need to cover thematic areas such as the role of indigenous tourism and desert tourism is a critical concern, given that Southern Africa is still home to the Khoi and San communities. These communities are largely domiciled in rural environments near some rural tourism hubs such as Ai-Ais Richtersveld National Park, Kgalagadi National Park, and the Kalahari desert areas. How tourism development has affected such indigenous populations is an area of concern. Moreover, there have been clashes in the Serengeti National Park with the Maasai Mara people (the indigenous people) (Goldman, 2011), who are threatened by tourism development and mass tourism in that region. Informed by the sustainable development goals, this particular interest in the tourism sector has been instrumental in advancing sustainable development, preserving biodiversity, and granting economic opportunities for local communities.

The slow growth in publications in the space of rural tourism is a threat to the achievement of Sustainable Development Goals in many respects, as it starves the policymakers and planners of much-needed data to make science-based decisions. Publications on this topic have been largely in the form of journal articles, which accounted for 75% (Fig. 10.2) of the published material. The second-most publications came in the form of book chapters, and only 3% came in the form of books. Only three reviews have been conducted on this topic, and conferences which act as wider general information dissemination have equally not been forthcoming regarding covering this topic. It might be crucial to have conferences that focus specifically on rural and protected area

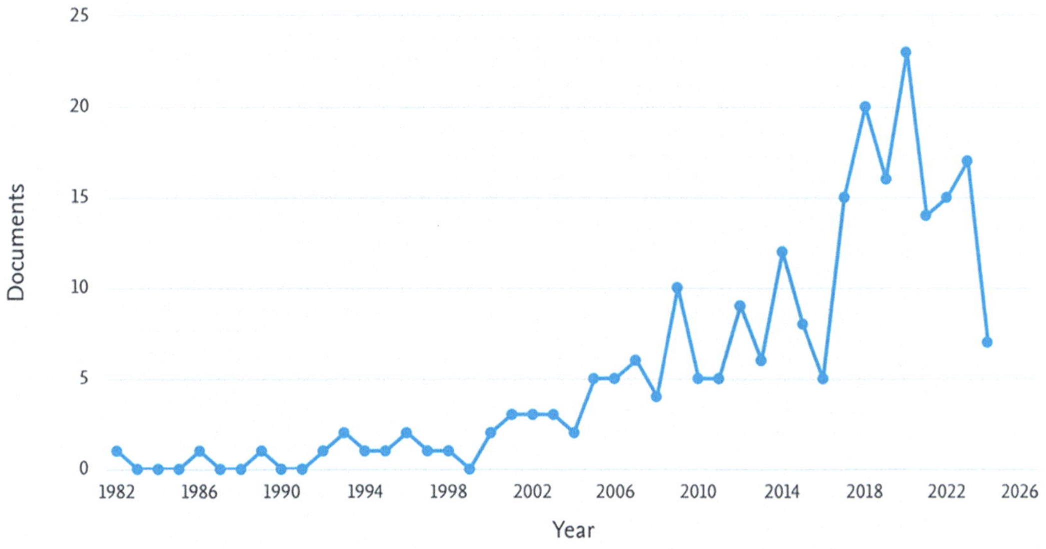

Fig. 10.1 Number of publications on rural tourism 1980–2024 (Source: authors' data from Scopus)

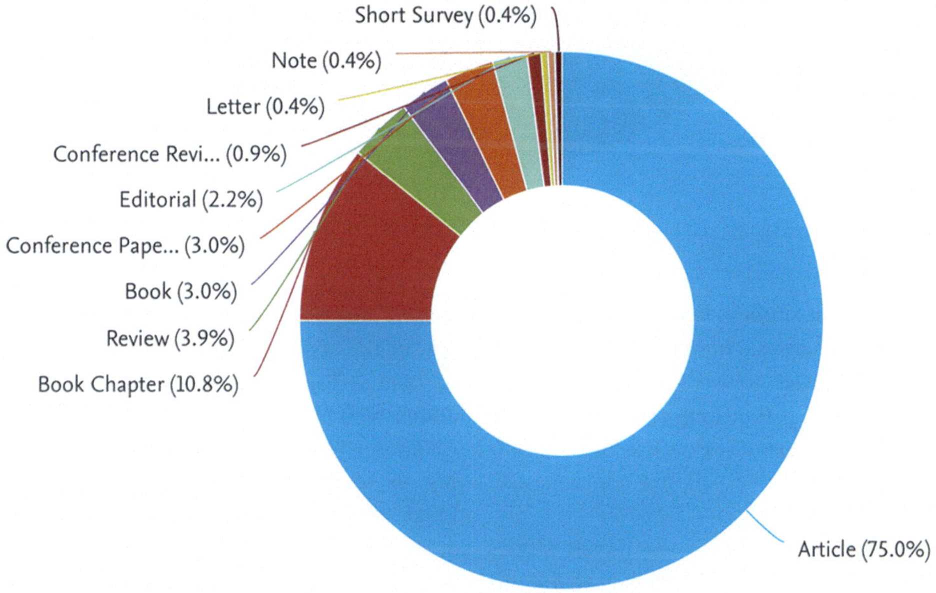

Fig. 10.2 Types of publications on rural tourism. (Source: authors' data from Scopus)

tourism to promote the discussions and development of this niche area as an area of focus.

The African Journal of Hospitality Tourism and Leisure has been at the forefront of publishing material on rural tourism in Southern Africa. The University of Johannesburg hosts the Journal, one of the biggest outlets for tourism and hospitality studies. The second biggest outlet for rural tourism studies is the *Journal of Development Southern Africa*, an international journal produced locally. Another critical outlet is the *South African Geographical Journal*, a locally produced international journal. This points out the significance and role of local journals in disseminating information on rural tourism. Given the transdisciplinary nature of the journal outlets, it

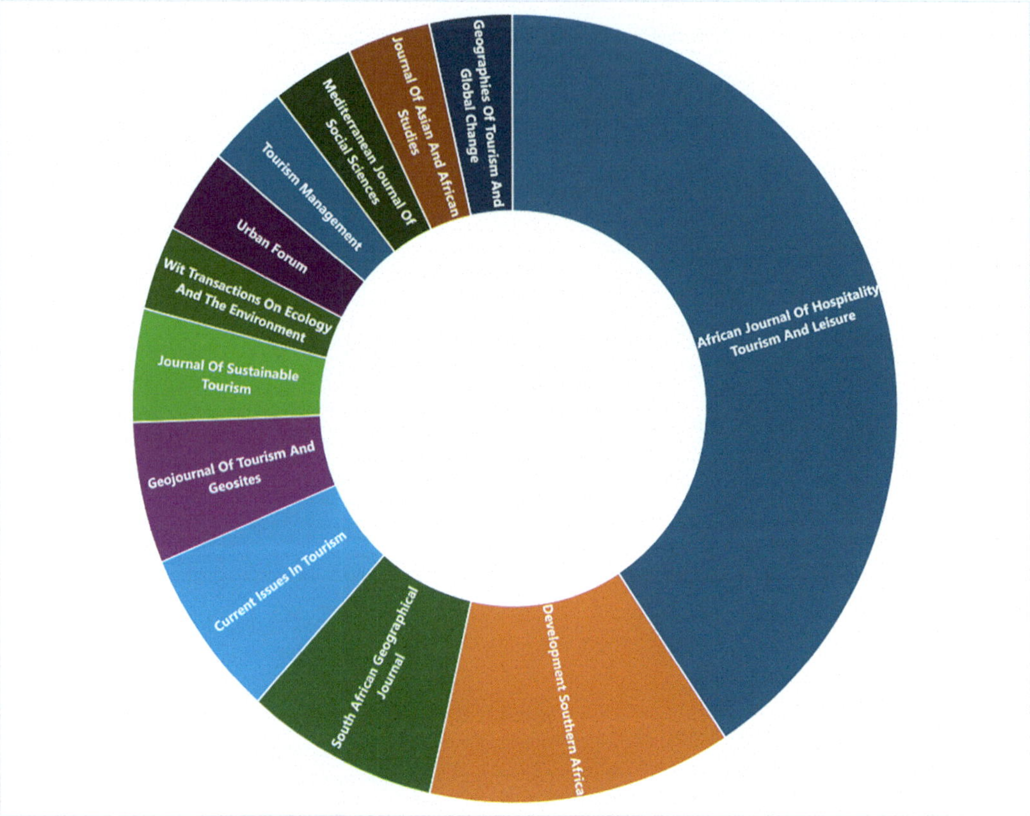

Fig. 10.3 Where research on rural tourism in Southern Africa. (Source: authors' data from Scopus)

is anticipated that studies on rural and protected areas tourism. There is a high number of leading international top-tier journals in Q1 that have covered material on this subject, which reflects that the research conducted on this topic in this area is of local and international relevance and appeal (Fig. 10.3).

Few scholars have made significant strides and focused approaches in publishing in this space, with Rogerson dominating this space. Rogerson CM and Rogerson J. M have contributed the highest number of studies in this space (Fig. 10.4). The two are accomplished tourism geographers who have contributed immensely to this space. Amongst others, Rogerson explored the historical evolution of the rural tourism landscape in South Africa (Rogerson & Rogerson, 2021b). Their other study Rogerson and Rogerson (2021a) called for a rethink of the policy terrain to ensure the enablement of rural tourism enterprises. They also explored the role of rural tourism in the socioeconomic development of rural tourism (Rogerson, 2015). Another notable author who made a dent in this space is Ramaano (2023a), who examined the usage of remote sensing in rural tourism sustainable development. Other studies explore nature tourism in rural areas (Ramaano, 2024) while examining the environmental impact of rural tourism establishments (Ramaano, 2023b).

10.2 Thematic Issues Covered in Rural Tourism

There are various areas of focus that have been chased by research scholars in rural tourism terrain, as can be seen in Fig. 10.5. There has been an evolution of focus areas for rural tourism in Southern Africa. Previously, the major focus of

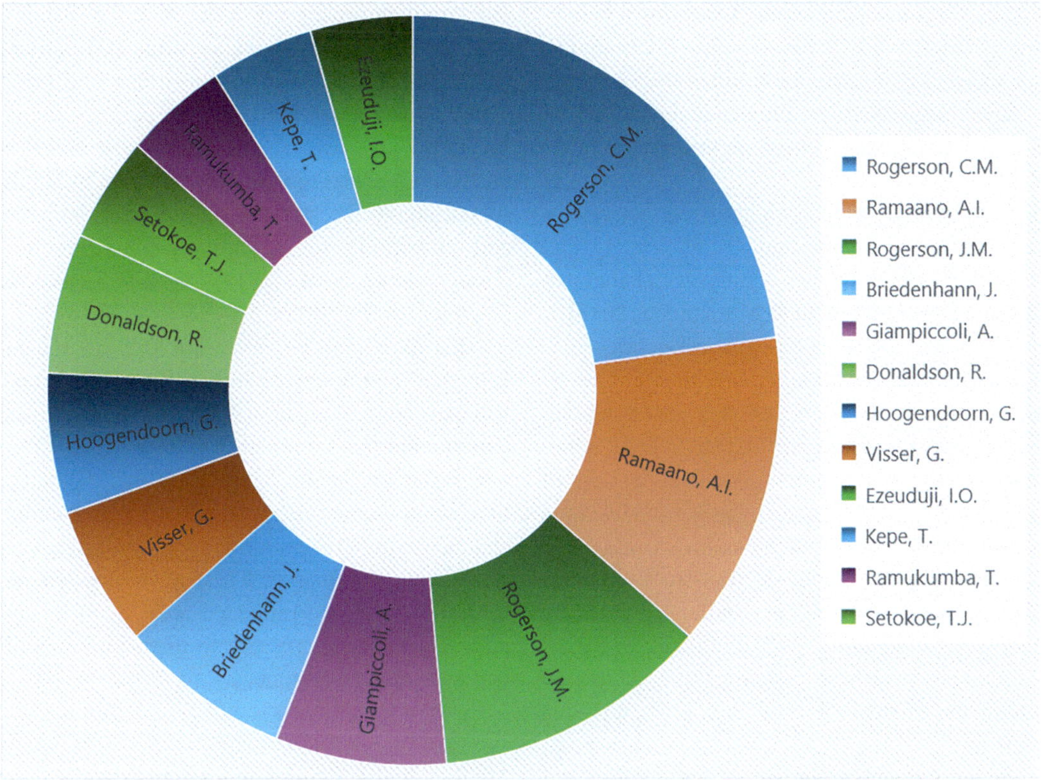

Fig. 10.4 Top scholars on rural tourism in South Africa. (Source: authors' data from Scopus)

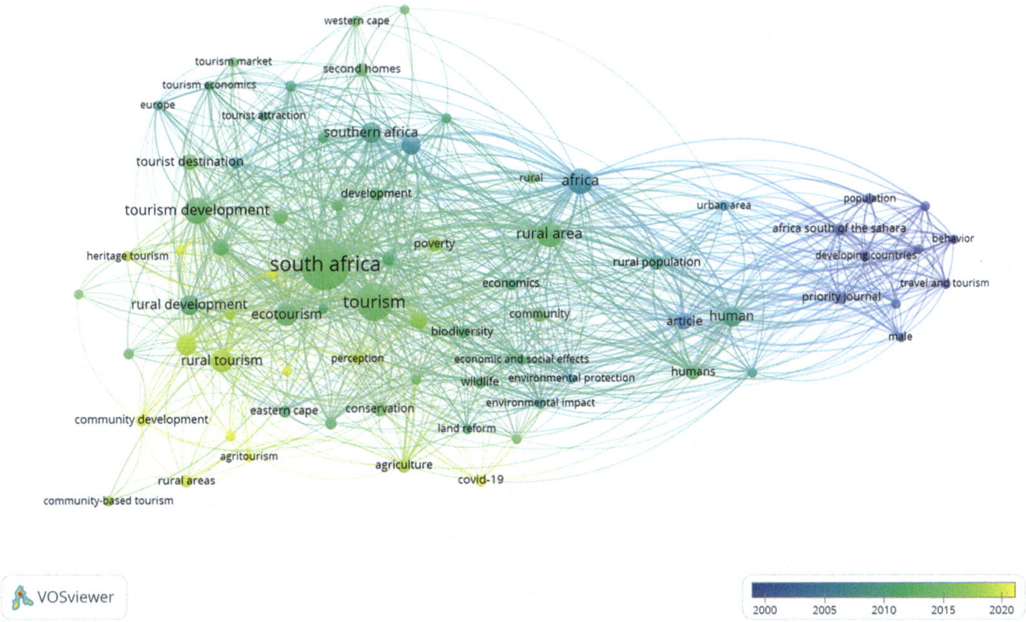

Fig. 10.5 Evolving themes in rural tourism in Southern Africa: (Source: authors' data from Scopus 1980–2024)

studies was on understanding the tourism behaviour of rural tourism participants. Such studies aim to serve that particular tourism market better by understanding its profile needs and wants. In some spaces, there has been an interest in understanding the role of rural tourism in community development. Rural tourism has been viewed as a critical driver of addressing rural poverty and economic development catalysts (Jamal & Dredge, 2014; Mthembu & Mutambara, 2018).

The issues of rural tourism, economic community development, and environmental conservation are at the core of this development. Addressing the economic developmental challenges facing rural communities is at the core of addressing SDG 1, addressing poverty. Given the region's high poverty and inequality levels, most host communities count on tourism to address some of the socioeconomic challenges faced by people in the region. As such, the issue of poverty is cited as a critical thematic area addressed within the context of rural and protected area tourism.

Environmental conservation issues are critical in addressing the aspirations set out in SDG 13 on climate change action, SDG 14, life below water, and SDG 15 on land. This is in light of the biodiversity threat of climate change in terrestrial and coastal areas (Mgadle et al., 2022; Sibitane et al., 2022). The other challenges facing rural and peri-urban tourism now extend to the threat caused by pollution due to poor municipal management and governance.

There has equally been a fair share of publications that focus on heritage tourism. This underscores the centrality of heritage areas in shaping tourism development in South Africa. South Africa is home to 12 properties listed on the World Heritage List. Most of these properties are located in various parts of the country. In addition, several other provincial heritage sites and strategic heritage areas are critical to heritage tourism development. Besides South Africa, the Southern African region has several critical world heritage sites, which are critical tourism assets.

Other emerging yet interesting themes from the rural tourism nexus manifest in the mould of ecotourism and agritourism. This tourism typology is not surprising given the need for sustainable tourism development, as discussed earlier. Given the vast farming in peri-urban and rural environments, the emergence of agritourism is not surprising. Several game farms that contribute to tourism and tourism development in South Africa are in the business of breeding animals and embarking on several tourism activities (Hoogendoorn et al., 2019; Van der Merwe & Saayman, 2003, 2005).

That as it may be, there are still significant gaps that require attention; vast knowledge gaps still exist concerning how, for example, how rural tourism interfaces with other SDGs, such as SDG 6 on water and sanitation, SDG 5 Gender quality, SDG 4 on quality education, SDG 3 on healthy and the nexus between rural tourism and issues of clean energy production. These are issues that are central to rural and protected areas as these are matters which directly talk to the aspirations of that particular constituency.

10.3 Book Chapter Summaries

Chapter 1 reflects on the importance of ecotourism. Lately, scholars favour referring to the broader concept of *sustainable tourism* in studies on rural tourism and protected areas, which are closely linked to sustainable development goals (Gallagher, 2021). Scholars Nhamo et al. (2023) and Ezeuduji and Mgabhi (2023) have highlighted the importance of rural tourism as a means for socioeconomic development, especially in protected areas that are impacted by climate change. Dube and Nhamo (2020) and Dube et al. (2022) have emphasised the role of protected areas and how climate change can impact conservation efforts and hamper sustainable development.

Based on the journal titles shown in Fig. 10.4, the relationship between protected areas and conservation is particularly interesting to the Southern African tourism sector. With renowned sites like the Kruger National Park, Hluhluwe-iMfolozi Park, Kgalagadi Transfrontier Park, and Okavango Delta, Southern Africa is home to an array of protected areas that serve as a foundation

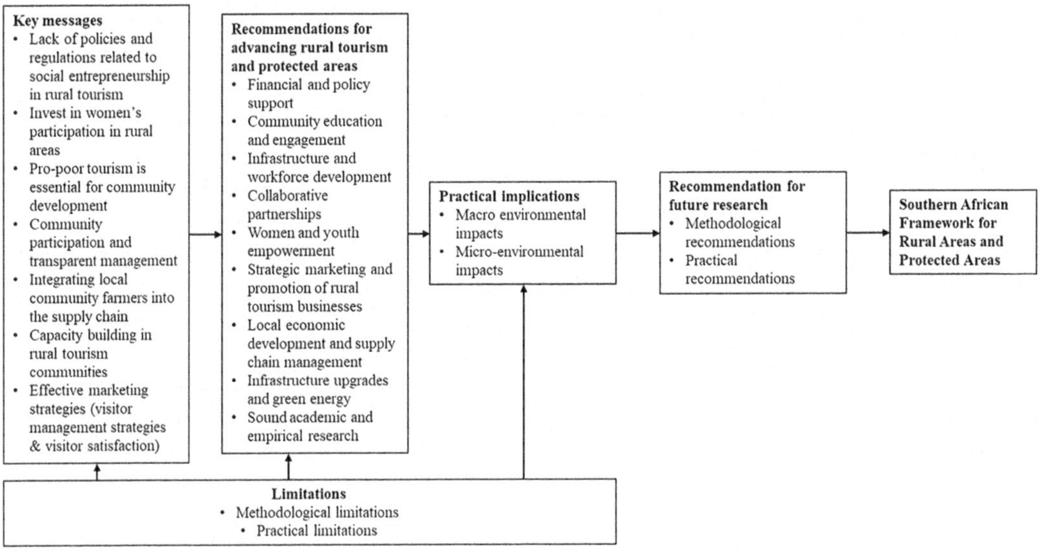

Fig. 10.6 Chapter outline. (Source: Authors' contribution)

of biodiversity and anchors rural tourism. Scholars in this field have advocated for more integrated community-focused tourism models (Mtapuri et al., 2021). This emerging trend aims to balance the interplay between conservation and community development, which resonates with all the chapter contributors throughout this collection.

This final chapter synthesised the key messages, recommendations for advancing rural tourism and protected areas, practical implications, limitations, and recommendations for future studies. We highlight the critical success factors for developing and conserving rural tourism in protected areas, the challenges to maintaining infrastructure and supply chains, and the importance of meeting evolving tourism expectations. This collective finding gives insights into new trends in sustainable marketing strategies, innovative tourism offerings aligned with sustainable development goals, and robust management practices to balance the need for enhanced tourism experiences while ensuring these protected areas' and communities' long-term viability. Figure 10.6 outlines the most relevant discussions from this chapter to inform the development of a Southern African Framework for Rural Tourism and Protected Areas.

Through the case studies and empirical research, this chapter contributes to a broader discourse on rural tourism and protected areas. It points out the need for a holistic approach to tourism conservation that integrates environmental sustainability, community engagement, capacity building, and economic development. The proposed recommendations aim to provide practical solutions to the current challenges for sustainable rural tourism practices that will make the protected areas sought-after tourism destinations. Therefore, this chapter critically reflects on the current state of tourism in Southern Africa and provides actionable insights for practitioners, policymakers, and scholars.

10.4 Key Messages

This chapter contributors underscored several critical messages to navigate the complexities of rural tourism and the management of protected areas. Challenges such as resource constraints, inadequate infrastructure, limited community support, and a **lack of policies and regulations related to social entrepreneurship in rural tourism** areas are among the most universal constraints (Sharifi-Tehrani et al., 2022. All key stakeholders (such as governments, social entrepreneurs, and community leaders) need an integrated approach to address these challenges, which include financial support, capacity building, and collaboration. Through these collective initiatives, infrastructure must be upgraded and developed, training and skills transfers to the local workforce must be prioritised, and coordination and communication between the social entrepreneurs and government agencies must be established.

Although **women's participation in rural areas** shows progress towards gender equality and economic progress, many are still marginalised and have limited access to business networks and financing (Swart et al., 2024). Gender disparities and women's restricted decision-making roles remain a significant challenge in rural areas. Effective gender mainstreaming, tailored financial support, and effective marketing strategies for maximum exposure of female-owned tourism businesses are critical for empowering women and youth in rural areas in the vicinity of protected areas.

Pro-poor tourism is essential for community development (Musavengane, 2018). This is demonstrated in areas adjacent to protected regions like Hwange National Park, where these initiatives alleviate poverty and foster community development. Through tourism, communities expect direct and indirect benefits such as creating jobs and the equal distribution of revenues generated in the protected area. Addressing the challenges associated with inadequate funding, insufficient benefits for low-income people, and human–wildlife conflicts is critical for the sustainability and effectiveness of pro-poor tourism initiatives.

The success of community-based tourism businesses and income-generating projects in protected areas depends on active **community participation and transparent management** (Chiutsi & Saarinen, 2017). This will ensure that the economic benefits stemming from tourism will reach the poorest members of society and provide a decent living, as advocated by Sustainable Development Goal #8.

Integrating local community farmers into the supply chain of luxury lodges can improve regional economic development and create more sustainable supply chains in rural areas. The supply of these fresh produce demand is essential for the mutual support of agriculture and tourism in protected areas (Pooe & Mahlangu, 2017).

Capacity building in rural tourism communities can boost local economies (Imbaya et al., 2019). A holistic approach that focuses on redeveloping and maintaining the ageing infrastructure of local attractions, cooperative governance, specialised training across the rural tourism value chain, and strong leadership is needed to integrate local resources and attractions into a rural tourism framework. A capable and empowered local workforce can develop effective **visitor management strategies** to attract visitors to rural areas (Leung et al., 2018). Infrastructure designs, eco-labels, education, and visitor codes of conduct are desired visitor management tools to promote ethical behaviour amongst all stakeholders.

Effective marketing strategies for protected areas, such as the Hluhluwe-iMfolozi Park, should include push factors (such as relaxation) and pull factors (such as natural resources). However, promotional materials must consider these factors to target the right market (Ezeuduji & Mgabhi, 2023). **Visitor satisfaction** is enhanced once the visitor and tourism expectations are aligned with the actual experiences to ensure quality in service delivery and visitor retention (Mutanga et al., 2017), especially in protected areas such as the Kruger National Park. Through these critical messages, tourism managers and policymakers can collaborate to have a sustainable impact on local communities and conservation efforts.

10.5 Recommendations for Advancing Rural Tourism and Protected Areas

Grounded by the research from the preceding chapters, rural tourism in protected areas offers unique opportunities to promote sustainable development, as evidenced by the following recommendations.

Financial and policy support is needed to enhance protected areas and the capacitating of the local community (Majenga & Mashenene, 2015). This can be done by establishing dedicated *funds and grant programmes* by governments and philanthropic organisations. Social entrepreneurs and community-based women-owned tourism businesses must receive priority funding to alleviate resource constraints and build businesses. The *bureaucratic processes* of conducting business must be simplified, supported by tax incentives, and a conducive legal environment in which companies are protected against exploitation. *Gender responsive policies* that address gender inequalities promote women's entrepreneurship and ensure equal access to resources and networks must be prioritised. This must include Gender Governance Systems like Gender Responsive Budgeting.

Community education and engagement are the cornerstone of sustainable development in rural areas (Tichaawa & Mhlanga, 2015). *Entrepreneurship awareness* must be created by developing and implementing community education programmes, outreach initiatives, and workshops to promote tourism stakeholder projects. Developing viable community-based tourism businesses owned and managed by locals can generate significant income and employment opportunities to address poverty in these areas, as advocated by Sustainable Development Goal #1.

Infrastructure and workforce development are custodians of local economic development (Rogerson, 2006). Government entities must prioritise investment and updates to infrastructure to ensure modern standards and enhanced visitor experience, especially in protected areas such as the Kruger National Park. This must be complemented by workforce training in collaboration with educational institutions. The development of tourism-related business management skills can be complemented by training programmes, internships, and apprenticeships to equip women and youth especially.

Collaborative partnerships focusing on *multi-stakeholder* and *public–private partnerships* can support and ensure the success of tourism businesses in rural areas (McCamley & Gilmore, 2017). Sharing resources and knowledge through partnerships between governments, local tourism businesses, non-profit organisations, private investors, and development agencies can capacitate developing tourism businesses to achieve Sustainable Development Goal #17 (partnerships for goals). Fostering public–private partnerships between protected areas (as in the case of Hluhluwe-iMfolozi Park) and neighbouring communities can ignite the development of innovative and effective tourism offerings.

Women and youth empowerment in tourism is critical to economic and social progress in rural tourism communities. *Access to finance* through tailored microloans and technology support can ensure that women are empowered with the necessary leadership skills to have rightful participation in the rural tourism economy. Women-owned and managed tourism businesses must be *recognised and promoted* in marketing materials as viable businesses to shift societal perceptions and cultural stereotyping often associated with women in tourism (Yang et al., 2024).

Various variables influence rural tourism businesses' **strategic marketing and promotion** (Wambani et al., 2020). This ranges from *visitor management and experience enhancement* to a *comprehensive visitor management strategy*. Destination managers and tourism product owners must implement marketing strategies focusing on specific target markets tailored to the protected areas' unique attractions. By aligning promotional efforts with the push and pull factors of each attraction and protected area, the visitor experience can be enhanced. Furthermore, by upgrading the on-site services, providing consistent pre-visit information and developing innovative and personalised tourism packages, the visitor experience can be met or exceeded. Visitor management tools can improve the integration of social norms into the visitor experience to promote ethical behaviour, especially in protected areas. First-time visitors from various demographic groups can be introduced to the different protocols and ethical traditions when visiting rural and protected areas.

Local economic development and supply chain management emphasise the importance of responsible tourism practices for community involvement and sustainable resource management (Van den Berg & Mearns, 2021). The authors recommend *identifying and supporting local suppliers* and creating a *marketplace platform* to promote local produce to tourism businesses in rural and protected areas. Opportunities have emerged for luxury wildlife lodges, especially in collaboration with local community farmers, to expand the variety of fresh produce available. Through such initiatives, tourists are introduced to the gastronomy of the local area and educated about the community's cultural heritage through food. This symbiotic relationship between tourism businesses and local agriculture is aligned with Sustainable Development Goal #12 (responsible consumption and production). Social media can market and promote local produce and curios by making them accessible to the broader public, communities, and lodges to foster local economic development.

Infrastructure upgrades and green energy use are effective ways to take climate action (Sustainable Development Goal 13) in rural and protected areas. Accelerating the *adoption of renewable energy* within protected areas, such as solar installation, can make the areas more energy-secure and ensure that business operations are environmentally sustainable (Osman et al., 2023). Furthermore, *upgrading outdated infrastructure*, especially in protected areas, through incorporating green building designs can enhance tourism businesses' economic and environmental sustainability.

None of these recommendations are actionable if this is not complemented with sound **academic and empirical research**. It is, therefore, necessary to expand our research efforts on visitor management, financial and policy support, community education and engagement, infrastructure and workforce development, collaborative partnerships, women and youth empowerment, strategic management, local economic development, and green energy initiatives. Research is further dominated by male research, as illustrated in Fig. 10.6.

As men dominate most research on rural tourism and protected areas, it is necessary to capacitate young women researchers for a balanced approach when interpreting their research findings to make this actionable for policy implementation.

10.6 Practical Implications for Rural Tourism and Protected Areas

The studies in this collection guide the practical implications, which aim to inform tourism stakeholders on supporting rural tourism development and capacitating the conservation of protected areas for future generations. These implications are presented against the macro- and micro-environments that may impact a rural or tourism business in a protected area.

Micro-environmental impacts entail the strengths and weaknesses of the rural tourism business when they engage with the tourists (customers), competitors, intermediaries, general public, and suppliers of tourism products (Nel & De Beer, 2018). Therefore, *mobilising resources* and developing *financial strategies* highlighted the importance of diversified funding to overcome resource constraints. Strategies that support the economic independence of women and youth in tourism will enable them to explore new market opportunities in rural areas and invest in tourism businesses. Stokvels have been identified as a variable tourism investment tool that can help governments overcome resource constraints and support sustainable community development (Seoe & Swart, 2023). *Customised visitor management tools* need to focus on social norms and refined strategies to address a specific demographic behaviour; for example, one of the iSimangaliso Wetland Park (South Africa) strategies is to focus on repeat visitors and married couples. This will *enhance visitor satisfaction* as the marketing strategies align with the tourists' expectations and motivations. For example, at the

Hluhluwe-iMfolozi Park, improved visitor satisfaction can stimulate repeat visits and positive word-of-mouth, resulting in the long-term success of this rural tourism destination in a protected area. Other protected areas, such as Kruger National Park, must *prioritise infrastructure upgrades* by investing in green energy and modern facilities to meet the evolving needs of tourists (customers) and to position themselves as leaders in sustainable tourism. Implementing *advanced property management systems*, staff training, and supervision will improve operational efficiency while reducing service failures and strengthening the reputation of the protected area as a premier tourist destination. Local communities in rural areas are empowered by *education and community engagement* initiatives. When communities actively participate in tourism-related activities, it strengthens the community bond and ensures that benefits are equally distributed, primarily through Pro-poor Tourism initiatives. Education promotes an understanding and appreciation for social entrepreneurship (Sharifi-Tehrani et al., 2022), which fosters long-term growth to benefit nature and local populations.

Macro-environmental impacts involve tourism opportunities and threats that may emerge from the economic-, political-, demographic-, socio-cultural-, technological-, and ecological environments (Nel & De Beer, 2018). *Collaborating networks* and *technology integration* can help tourism entrepreneurs reach new audiences. Networking and collaboration among government agencies, local businesses, and social entrepreneurs support the development of best practices for tourism in rural areas and a culture of cooperation. This particularly interests women in tourism associations and mentorship programmes supporting the expansion of women-led businesses. Tourism entrepreneurs must embrace technology and digital tools to expand their business, reach new target markets and strengthen their roles in the industry. Technology education is of specific interest to women and contributes to gender equality and sustainable development in rural and developing areas. *Cultural and environmental stewardship* is needed to address the challenges of conservation. Tourism stakeholders must adopt sustainable tourism practices, such as integrating income-generating projects (such as livestock farming or beekeeping) in rural tourism businesses. Diversifying the local rural economy will preserve the area's cultural heritage and encourage the conservation of natural resources in the protected areas. The tourism community can overcome challenges like equitable benefit contribution, funding shortfalls, and human–wildlife conflicts, ensuring long-term sustainable initiatives in rural and protected areas through stakeholder collaboration. *Strategic supply chain management* at luxury wildlife lodges where local farmers are integrated into the supply chain will support local economies, enhance sustainability, and limit the disruptions in delivering goods and services to the protected area. This collaboration with local farmers can improve the variety and quality of products while enriching the culinary experiences for tourists. More *local markets* are created as local artists can sell locally produced handcrafts and other local curios to tourists, directly supporting the local economy and offering unique and authentic experiences.

Responsible tourism practices can benefit from these micro- and macro-environmental impacts, involving both the local community and visitors (tourists). These impacts align the rural tourism initiatives with *global sustainability agendas*, such as the 2030 African Union Agenda and the United Nations Sustainable Development Goals. These strategies and findings from the studies on protected areas and tourism can be adapted and replicated in other regions across Africa and globally, contributing to tourism development frameworks that can inform global best practices.

10.7 Limitations

Throughout the collection, the authors have highlighted several limitations. These are clustered as methodological and practical limitations which aim to inform similar studies on rural tourism and protected areas. Several **methodological limitations** were highlighted, ranging from resource constraints to data management. *Insufficient data collection methods*, *unethical data collection practices*, and an *overreliance on third-party data* compromise the long-term sustainability of research projects in rural areas. Incomplete or irrelevant data impacts the accuracy and comprehensiveness of reliable findings and conclusions. Potential *bias in self-reported data* may affect the accuracy and reliability of findings. Researchers must caution against overreliance on only tourist feedback as this may exclude the perspectives of other stakeholders concerned about managing rural tourism resources.

Most of the identified limitations are from a **practical perspective**. Access to sufficient *human* (talent) and *financial resources* is among the most universal challenges in developing rural tourism and protected areas. Most projects in these areas are often over-reliant on government grants, philanthropic support, and unreliable external funding, which hampers the development of sustainable tourism initiatives. Rural areas are known to have *slow infrastructure development*, which is further delayed by budgetary constraints and unreliable public–private partnerships. The complexities associated with infrastructure building and the consistent need for investment in remote areas are ongoing challenges. *Bureaucratic barriers* and inconsistent or slow policy implementation create significant obstacles. *Complex regulatory environments* create a gap between rural tourism and protected area policy implementation and practical application in communities, delaying the effectiveness of the intended tourism strategies. Resistance, cultural beliefs, and low literacy levels often hamper *educational awareness* campaigns in rural communities. This calls for tailored community engagement projects that incorporate tourism practices and are communicated in a learning-friendly environment. However, increasing participation in these capacity building projects does not guarantee success and requires a *well-integrated project plan* where all tourism stakeholders can provide inputs. Diverse stakeholder interests *hamper collaborative* efforts to develop effective tourism partnerships among local communities, the private sector and the government, especially when the rural development goals and priorities are diverse. Due to the *geographical isolation* of rural areas and protected areas, there is consistent competition for resources and networking opportunities, especially for women and other marginalised groups. As social networks mostly revolve around men, women's ability to participate in tourism economic activities is severely hampered by these social sanctions (Morgan & Winkler, 2020).

The *scalability* of the application of findings on the advancement of rural tourism may not be relevant to other regions due to the community's cultural context and socioeconomic perspectives, which limits the monitoring of success strategy implementation over time. Insights are also location-specific or presented from an African epistemology, which may not universally appeal to rural areas in other developing countries and require more context-specific research. The *monitoring and implementation* of rural tourism initiatives require continuous support and adaptability by all tourism stakeholders, which requires substantial resources which are often exhausted before the full impact can be determined.

Besides conservation efforts and investment in research, implementing strategies to mitigate *human–wildlife conflict* continues to hamper the effectiveness of conservation efforts. *Resource disparities* related to access to training opportunities and financial resources for women and youth in rural communities continue to constrain the effectiveness of tourism entrepreneurial opportunities (Tshabalala & Ezeuduji, 2016). The *digital divide* further enhances these limitations, where marginalised groups (women and youth) face barriers to utilising digital technology (Figueroa-Domecq et al., 2020).

10.8 Conclusion and Recommendations for Future Research for Rural Tourism and Protected Areas.

Future research can provide a more nuanced, practical, and sustainable approach to rural tourism and protected areas.

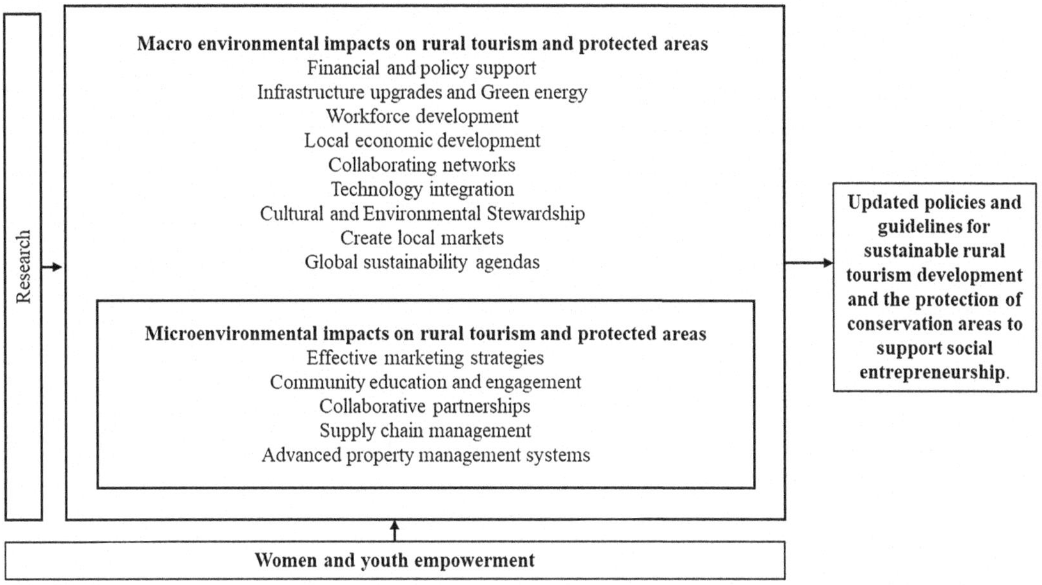

Fig. 10.7 Southern African framework for rural areas and protected areas. (Source: Authors' contribution)

From a **methodological perspective**, *longitudinal impact assessments* are needed to assess the long-term impact of tourism development in rural areas and related protected areas, especially in the capitating of women and youth as entrepreneurs. This will provide an in-depth understanding of the evolution of sustainable practices of tourism projects over time and its impact on the environment. Furthermore, a more holistic approach incorporating *different methodologies*, where external factors are considered and how this impacts long-term sustainability, is needed. Through mixed methods approaches, researchers can engage with various stakeholders (respondents and participants) to address multiple dimensions of tourism development, leading to more adaptable and robust solutions for rural tourism and protected areas.

Practical recommendations for future research entail *implementing the designed strategies* and a *comprehensive resource assessment* when tourism strategies are developed. A thorough examination of local resources, such as financial, human, and talent resources and infrastructure, will support improving tourism strategies. Practical guidance is needed on overcoming the identified challenges in rural areas and which tourism strategies are required to address these challenges effectively. This calls for developing relevant, actionable, sustainable tourism frameworks and where the recommendation can be implemented. These future research initiatives can improve local communities' outcomes, especially those near protected areas, and our conservation efforts. Based on these recommendations, there is an opportunity to empirically validate and refine the proposed Southern African Framework for Rural Areas and Protected Areas, which is informed by the discussions from this collection, as illustrated in Fig. 10.7.

Prominent themes from the eight-chapter contributions were clustered under macro- and micro-environmental impacts on rural tourism and protected areas, as illustrated in Fig. 10.7. The *Southern African Framework for Rural Areas and Protected Areas* is of particular interest for empowering women and youth, as specific policies and guidelines are needed for economic and socio-cultural inclusion in rural areas and the management of protected areas.

References

Chiutsi, S., & Saarinen, J. (2017). Local participation in transfrontier tourism: Case of Sengwe community in Great Limpopo Transfrontier Conservation Area, Zimbabwe. *Development Southern Africa, 34*(3), 260–275. https://doi.org/10.1080/0376835X.2016.1259987

Dube, K. (2024). Evolving narratives in tourism and climate change research: Trends, gaps, and future directions. *Atmosphere, 15*(4), 4. https://doi.org/10.3390/atmos15040455

Dube, K., Chikodzi, D., Nhamo, G., & Chapungu, L. (2023). Climate and conservation challenges facing Marakele National Park and their implications for tourism. *Cogent Social Science, 9*(2), 2282705. https://doi.org/10.1080/23311886.2023.2282705

Dube, K., Kupika, O. L., & Chikodzi, D. (2023). *COVID-19, tourist destinations and prospects for recovery: Volume three: A South African and Zimbabwean perspective*. Springer International Publishing. https://doi.org/10.1007/978-3-031-28340-6

Dube, K., & Nhamo, G. (2020). Evidence and impact of climate change on South African national parks. Potential implications for tourism in the Kruger National Park. *Environmental Development, 33*, 100485. https://doi.org/10.1016/j.envdev.2019.100485

Dube, K., Nhamo, G., & Chikodzi, D. (2022). Climate change-induced droughts and tourism: Impacts and responses of Western Cape province, South Africa. *Journal of Outdoor Recreation and Tourism, 39*, 100319. https://doi.org/10.1016/j.jort.2020.100319

Ezeuduji, I. O., & Mgabhi, N. Z. (2023). Potential for rural tourism development around Hluhluwe iMfolozi Park, South Africa: A local community perspective. *African Journal of Development Studies, 13*(1), 89–113. https://doi.org/10.31920/2634-3649/2023/v13n1a5

Ferreira, S. L. A. (1999). Crime: A threat to tourism in South Africa. *Tourism Geographies, 1*(3), 313–324. https://doi.org/10.1080/14616689908721324

Figueroa-Domecq, C. F., Williams, A., de Jong, A. D., & Alonso, A. (2020). Technology is a woman's best friend: Entrepreneurship and management in tourism. *e-Review of Tourism Research, 17*(5), 777–792. https://journals.tdl.org/ertr/index.php/ertr/article/view/560

Gallagher, K. (2021). *Sustainable tourism vs. ecotourism: What's the difference?* Retrieved from https://www.tripsavvy.com/sustainable-tourism-ecotourism-differences-5185311

Goldman, M. J. (2011). Strangers in their own land: Maasai and wildlife conservation in Northern Tanzania. *Conservation and Society, 9*(1), 65. https://doi.org/10.4103/0972-4923.79194

Hoogendoorn, G., Meintjes, D., Kelso, C., & Fitchett, J. (2019). Tourism as an incentive for rewilding: The conversion from cattle to game farms in Limpopo province, South Africa. *Journal of Ecotourism, 18*(4), 309–315. https://doi.org/10.1080/14724049.2018.1502297

Imbaya, B. O., Nthiga, R. W., Sitati, N. W., & Lenaiyasa, P. (2019). Capacity building for inclusive growth in community-based tourism initiatives in Kenya. *Tourism Management Perspectives, 30*, 11–18. https://doi.org/10.1016/j.tmp.2019.01.003

Ingle, M. (2010). Making the most of 'nothing': Astro-tourism, the sublime, and the Karoo as a 'space destination'. *Transformation: Critical Perspectives on Southern Africa, 74*(1), 87–111. https://muse.jhu.edu/pub/99/article/444006/summary?casa_token=rDwfs7SnU-wAAAAA:Zm7U3VKDAQz5ZbiYYz9DdEoHo1UEESh691EsPcMFpzZD5AWlbV9EL100cnroJxu43aBIXngulg

Jacobs, L., Du Preez, E. A., & Fairer-Wessels, F. (2020). To wish upon a star: Exploring astro tourism as vehicle for sustainable rural development. *Development Southern Africa, 37*(1), 87–104. https://doi.org/10.1080/0376835X.2019.1609908

Jamal, T., & Dredge, D. (2014). Tourism and community development issues. In R. Sharpley & D. J. Telfer (Eds.), *Tourism and development: Concepts and issues* (pp. 178–204). Channel View Publications. https://doi.org/10.21832/9781845414740-008

Jiwaji, N. T. (2016). Astro-tourism as a high potential alternative tourist attraction in Tanzania. *Huria: Journal of the Open University of Tanzania, 23*(1), 106–113. https://www.ajol.info/index.php/huria/article/view/152728

Leung, Y., Spenceley, A., Hvenegaard, G., Buckley, R., & Groves, C. (2018). *Tourism and visitor management in protected areas: Guidelines for sustainability*. IUCN.

Majenga, A., & Mashenene, R. (2015). Socio-cultural factors and financial performance among women small and medium enterprises in Tanzania. *European Journal of Business Management, 6*(32), 153–163.

McCamley, C., & Gilmore, A. (2017). Aggravated fragmentation: A case study of SME behaviour in two emerging heritage tourism regions. *Tourism Management, 60*(1), 81–91.

Mgadle, A., Dube, K., & Lekaota, L. (2022). Conservation and sustainability of coastal city tourism in the advent of seal level rise in Durban, South Africa. *Tourism in Marine Environments, 17*(3), 179–196. https://doi.org/10.3727/154427322X16599399156575

Minnaar, A., & Herbig, F. (2018). The impact of conservation crime on the South African rural economy: A case study of rhino poaching. *Acta Criminologica: African Journal of Criminology & Victimology, 31*(4), 147–168. https://doi.org/10.10520/EJC-159791abf8

Morgan, M. S., & Winkler, R. L. (2020). The third shift? Gender and empowerment in a women's ecotourism cooperative. *Rural Sociology, 85*(1), 137–164. https://doi.org/10.1111/ruso.12275

Mtapuri, O., Camilleri, M. A., & Dluzewska, A. (2021). Advancing community-based tourism approaches for the sustainable development of destinations. *Sustainable Development, 30*(3), 423–432.

Mthembu, B., & Mutambara, E. (2018). Rural tourism as a mechanism for poverty alleviation in Kwa-Zulu-Natal Province of South Africa: Case of Bergville. *African Journal of Hospitality, Tourism and Leisure, 7*(4), 1–22. https://www.ajhtl.com/uploads/7/1/6/3/7163688/article_58_vol_7_4__2018.pdf

Musavengane, R. (2018). Toward pro-poor local economic development in Zimbabwe: The role of pro-poor tourism. *African Journal of Hospitality, Tourism and Leisure, 7*(1), 1–14.

Mutanga, C. N., Vengesayi, S., Chikuta, O., Muboko, N., & Gandiwa, E. (2017). Travel motivation and tourist satisfaction with wildlife tourism experiences in Gonarezhou and Matusadona National Parks, Zimbabwe. *Journal of Outdoor Recreation and Tourism, 20*, 1–18. https://doi.org/10.1016/j.jort.2017.08.001

Nel, J., & De Beer, A. (2018). *Business management* (3rd ed.). Juta Inc.

Nhamo, G., Dube, K., Chapungu, L., & Chikodzi, D. (2023). Quest for NetZero emissions in South African national parks: A tourism perspective. *Heliyon, 9*(6), e16410. https://doi.org/10.1016/j.heliyon.2023.e16410

Ntuli, T. G., & Potgieter, P. J. (2001). Exploring the Impact of crime on tourism in St. Lucia. *Acta Criminologica: African Journal of Criminology & Victimology, 14*(1), 60–70. https://doi.org/10.10520/EJC28668

Osman, A. I., Chen, L., Yang, M., Msigwa, G., Farghali, M., Fawzy, S., Rooney, D. W., & Yap, P.-S. (2023). Cost, environmental impact, and resilience of renewable energy under a changing climate: A review. *Environmental Chemistry Letters, 21*(2), 741–764. https://doi.org/10.1007/s10311-022-01532-8

Pooe, R. I. D., & Mahlangu, D. (2017). Enhancing SME performance through supply chain integration, collaborative planning, and supply chain capabilities – research. *Journal of Contemporary Management, 14*(1), 238–269.

Ramaano, A. I. (2023a). Geographical information systems in sustainable rural tourism and local community empowerment: A natural resources management appraisal for Musina Municipality' Society. *Local Development & Society, 4*(1), 74–105. https://doi.org/10.1080/26883597.2021.2011610

Ramaano, A. I. (2023b). Nature and impacts of tourism development facilities and activities on the livelihoods of communities in Musina municipality. *Tourism Planning & Development, 20*(4), 696–720. https://doi.org/10.1080/21568316.2022.2115124

Ramaano, A. I. (2024). Environmental change impacts and inclusive rural tourism development on the livelihoods of native societies: Evidence from Musina Municipality, South Africa. *International Journal of Ethics and Systems, 40*(3), 495–525. https://www.emerald.com/insight/content/doi/10.1108/IJOES-04-2023-0089/full/html?casa_token=fxKeIKV9l4QAAAAA:fLVWb3p5Pk5kMGDK2WUJUehQ3e-OuGxDnvalRtr_9S06aLGfRIeSU2TbHjSXQBIvXav7a0eWBk2fZYHMN7Ez4jKN-HpBOq6tiEc3ETgCZIDf2aWmd2Q

Rogerson, C. M. (2006). Pro-poor local economic development in South Africa: The role of pro-poor tourism. *Local Environment, 11*(1), 37–60. https://doi.org/10.1080/13549830500396149

Rogerson, C. M. (2015). Tourism and regional development: The case of South Africa's distressed areas. *Development Southern Africa, 32*(3), 277–291. https://doi.org/10.1080/0376835X.2015.1010713

Rogerson, C. M., & Rogerson, J. M. (2021a). In-between' rural tourism: Rethinking rural tourism spaces for policy development in South Africa. *African Journal of Hospitality, Tourism and Leisure, 10*(4), 1394–1413. https://www.ajhtl.com/uploads/7/1/6/3/7163688/article_proof_18_10_4_1394-1413-2.pdf

Rogerson, C. M., & Rogerson, J. M. (2021b). The evolution of rural tourism in South Africa: An historical lens. *African Journal of Hospitality, Tourism and Leisure, 10*(6), 1728–1740. https://www.ajhtl.com/uploads/7/1/6/3/7163688/article_2_10_6_1728-1740.pdf

Seoe, N., & Swart, M. P. (2023). Putting money together: How tourism stokvels can improve living standards in South Africa. *Journal of Tourism & Development, 151*, 17242.

Sharifi-Tehrani, M., Seyfi, S., & Zaman, M. (2022). At the intersection of tourism social entrepreneurship and empathy: Development and validation of an empathy scale. *Journal of Business Research, 141*, 433–447.

Sibitane, Z. E., Dube, K., & Lekaota, L. (2022). Global warming and its implications on nature tourism at Phinda Private Game Reserve, South Africa. *International Journal of Environmental Research and Public Health, 19*(9), 9. https://doi.org/10.3390/ijerph19095487

Stone, C. (2019). *Prospects of an Astrotourism Site in Namibia*. https://digitalcommons.olivet.edu/pence_boyce/12/

Swart, M. P., Bernauer, V. S., & Thirumaran, K. (2024). Women's entrepreneurship in the hospitality and tourism industry: A systematic literature review and future research directions. In *Routledge handbook on gender in tourism: Views on teaching, research and praxis*. Taylor and Francis Group.

Tan, Y. H. M., & Nguwi, Y. Y. (2023). Tourism analytics—The case for South Africa. In Y. Y. Nguwi (Ed.), *Tourism analytics before and after COVID-19* (pp. 87–96). Springer Nature. https://doi.org/10.1007/978-981-19-9369-5_6

Tichaawa, T. M., & Mhlanga, O. (2015). Community perceptions of a community-based tourism project: a case study of the CAMPFIRE programme in Zimbabwe. *African Journal for Physical, Health Education, Recreation and Dance, 21*(Suppl. 1), 55–67. http://reference.sabinet.co.za/document/EJC182968

Tshabalala, S. P., & Ezeuduji, I. O. (2016). Women tourism entrepreneurs in KwaZulu-Natal, South Africa: Any way forward? *Acta Universitatis Danubius: Oeconomica, 12*(5), 19–32.

Van den Berg, H., & Mearns, K. (2021). An exploration of guesthouse fresh produce purchasing behaviour and supply chains in Johannesburg. *Journal of Transport and Supply Chain Management, 15*, 1–8.

Van der Merwe, P., & Saayman, M. (2003). Determining the economic value of game farm tourism. *Koedoe, 46*(2), 103–112. http://koedoe.co.za/index.php/koedoe/article/view/59

Van der Merwe, P., & Saayman, M. (2005). Game farms as sustainable ecotourist attractions. *Koedoe, 48*(2), 1–10. http://koedoe.co.za/index.php/koedoe/article/view/100

Wambani, P. P., Ogunjinmi, A. A., & Oladeji, S. O. (2020). Socio-demographic determinants of travel motivation and behaviour of visitors in nature-based destinations in northern Nigeria. *Journal of Applied Sciences and Environmental Management, 24*(12), 2121–2129. https://doi.org/10.4314/jasem.v24i12.18

Wassenaar, A., & Coetzee, B. (2024). Global astrotourism initiatives and the applicability of their strengths, weaknesses, opportunities and threats to astrotourism in South African National Parks. *Journal of Outdoor Recreation and Tourism, 46*, 100766. https://www.sciencedirect.com/science/article/pii/S2213078024000343

Weltman, A., Bull, P., Camera, S., Kelley, K., Padmanabhan, H., Pritchard, J., Raccanelli, A., Riemer-Sørensen, S., Shao, L., Andrianomena, S., Athanassoula, E., Bacon, D., Barkana, R., Bertone, G., Bœhm, C., Bonvin, C., Bosma, A., Brüggen, M., Burigana, C., & Gaensler, B. M. (2020). Fundamental physics with the Square Kilometre Array. *Publications of the Astronomical Society of Australia, 37*, e002. https://doi.org/10.1017/pasa.2019.42

Yang, E. C. L., Kimbu, A. N., Cai, W., & Swart, M. P. (2024). We are breaking the glass ceiling, but can we shatter the concrete roof: What is next for gender studies in tourism? In M. P. Swart, W. Cai, E. C. L. Yang, & A. N. Kimbu (Eds.), *Routledge handbook on gender in tourism: Views on teaching, research and praxis*. Taylor and Francis Group.

Index

K. Dube et al. (eds.), *Tourism in Rural Protected Areas*, Sustainable Development Goals Series,
https://doi.org/10.1007/978-3-031-87333-1

The manufacturer's authorised representative in the EU is Springer Nature Customer Service Centre GmbH, Europaplatz 3, 69115 Heidelberg, Germany. If you have any concerns regarding our products, please contact ProductSafety@springernature.com

Printed and bound by CPI Group (UK) Ltd, Croydon, CR0 4YY
01/10/2026
02225980-0020